W0255845

Doz. Dr. Bernd Hofmann

Born in 1953 in Eibau. Studied Mathematics from 1972 to 1976. Received Dr. rer. nat. in 1980, Dr. sc. nat. in 1984. Scientific assistant from 1976 to 1985 at the Technical University of Karl-Marx-Stadt, 1980 listener at the International Mathematical Centre "Stefan Banach" of Warsaw. Since 1985 associate professor in numerical mathematics at the Technical University of Karl-Marx-Stadt.

Fields of Research: Inverse and ill-posed problems, numerical methods for regularization, integral equations of the first kind, identification and control in partial differential equations.

Hofmann, Bernd:
Regularization for applied inverse and ill-posed problems. -
1. Aufl. - Leipzig: BSB Teubner, 1986. - 196 S.
(Teubner-Texte zur Mathematik; 85)
NE: GT

ISBN 978-3-322-93035-4 ISBN 978-3-322-93034-7 (eBook)
DOI 10.1007/978-3-322-93034-7
ISSN 0138-502X

Ursprünglich erschienen bei BSB B. G. Teubner Verlagsgesellschaft, leipzig, 1986
1. Auflage
VLN 294-375/54/86 · LSV 1085
Lektor: Dr. rer. nat. Renate Müller

Gesamtherstellung: Typodruck Döbeln, Bereich Leisnig
Bestell-Nr. 666 340 8
01900

TEUBNER-TEXTE zur Mathematik · Band 85



---

Bernd Hofmann

# Regularization for Applied Inverse and Ill-Posed Problems

## A Numerical Approach

The book presents numerical methods for the solution of linear and nonlinear inverse and ill-posed problems arising in science and engineering. A review of the theory and recent developments in this wide class of problems involving identification and control is illustrated by a series of examples. Using a general optimization approach, the main ideas of regularization as a strategy for the stable solution of ill-posed and ill-conditioned problems are discussed systematically. The book also gives an insight into the use of deterministic and stochastic a priori information in regularization. If regularization applies to discretized inverse problems, integral equations of the first kind, inverse problems in partial differential equations and inverse eigenvalue problems can be treated in a unified manner. An extensive bibliography completes the book.

Das Buch präsentiert numerische Methoden zur Lösung von linearen und nichtlinearen inversen und nichtkorrekten Aufgaben aus Wissenschaft und Technik. Anhand von Beispielaufgaben wird eine Übersicht zur Theorie und zu neueren Entwicklungen in dieser großen Klasse von Identifikations- und Steuerproblemen gegeben. Die systematische Darstellung der grundlegenden Ideen der Regularisierungsmethode als einer Strategie zur stabilen Lösung nichtkorrekter und schlechtkonditionierter Aufgaben nutzt einen allgemeinen Optimierungszugang. Das Buch vermittelt Einsichten in die Möglichkeiten der Nutzung von deterministischen und stochastischen a priori Informationen bei der Regularisierung. Wird die Regularisierungsmethode auf diskretisierte inverse Aufgaben angewandt, so lassen sich Integralgleichungen erster Art, inverse Probleme bei partiellen Differentialgleichungen und inverse Eigenwertprobleme in einheitlicher Weise behandeln. Ein umfangreiches Literaturverzeichnis komplettiert den Text.

Ce livre présente des méthodes numériques pour la résolution des problèmes d'inverses linéaires et nou linéaires et non corrects dans les sciences et les techniques. En partant de problèmes pris comme exemples, on donne une vue d'ensemble de la théorie et des nouveaux développements dans cette grande classe de problèmes d'identification et de contrôle. En partant d'une conception générale des idées fondamentals de la régularisation comme stratégie d'optimisation, on discute systématiquement les solutions stables des problèmes malposés et mal conditionnés. Le livre fournit des indications sur les possibilités d'utilisation d'informations déterministes ou d'informations stochastiques a priori pour la régularisation. Si l'on applique la regularisation aux problèmes d'inverse discretisé, on peut traiter d'une manière unifiée les équations intégrales de première espèce les problèmes inverses pour les équations aux derivées partielles et les problèmes aux valuers propres inverses. Une bibliographie très étendue vient completer le texte.

Книга посвящена численным методам решения линейных и нелинейных обратных и некорректных задач науки и техники. Теория и новые достижения в исследовании широкого класса задач идентификации и управления иллюстрируются на примерах. Систематичное изложение основ метода регуляризации как стратегия для получения устойчивых решений некорректных и плохо обусловленных задач использует общий подход оптимизации. Книга даёт понятие о возможности использования детерминированной и стохастической априорной информации для регуляризации. Применение метода регуляризации для дискретизированных обратных задач позволяет осуществить единый подход к решению интегральных уравнений первого рода, обратных задач для дифференциальных уравнений в частных производных и обратных задач на собственные значения. Текст дополняет обширный обзор литературы.

## PREFACE

As we shall see in the coming chapters, the efficient numerical treatment of applied inverse and ill-posed problems is founded on results from many different branches of mathematics. Arithmetic and algebra, set theory, functional and numerical analysis, optimization and statistics contribute much to the development and justification of solution strategies. The quantity and quality of current mathematical research on inverse reasoning is almost overpowering. However, the obtained results successfully apply to real life problems of inverse nature only if the chances and limitations of mathematics are completely recognized. It is characteristic of the family of inverse tasks that the given data do not involve enough information in order to recover a wanted quantity. To be more precise, small perturbations in the data may lead to large changes in the solution. This expresses the ill-posedness phenomenon arising in all classes of inverse problems. Exploiting additional a priori information seems to be absolutely necessary to overcome the disadvantageous effects of ill-posedness. It is evident that the collection of observation data and additional a priori information is never a mathematical problem. On the one hand, physicists, biologists and engineers can only utilize all the collected data if they are well-versed in performing and interpreting mathematical and numerical methods of identification and control. This implies a growing need for the non-mathematician to make himself aware of at least some of the major ideas and techniques of regularization. On the other hand, a mathematician who wants to apply regularization as the basic general concept for solving inverse problems also requires special insights into occurring variants of corresponding mathematical modelling. It is to help advance knowledge and ability to solve inverse problems for mathematicians, scientists and engineers that this text has been written.

In the coming six chapters, we shall introduce the reader to the basic aspects of classification, mathematical representation, discretization and numerical solution of inverse and ill-posed problems. The idea of regularization based on deterministic or stochastic models runs through most parts of this text. Of necessity the discussions of inverse problems will be limited in comprehensiveness and depth by the extent of the present volume. Thus, there was no space for illustrating the strategies and procedures by computational results. As for numerical effects concerning applied linear and nonlinear inverse example problems, the reader is referred to the author's papers [198-214], [137-38], [98], [133], [190] and [275]. The present text tries to make available many auxiliary statements from functional analysis, linear algebra, statistics and optimization theory in the form of lemmas with or without proof. However, it fails to be completely self-contained since it uses the basic mathematical notation and well-known analytic and algebraic dependences without further explanation. A commented list of notation at the end of the text should facilitate the understanding.

The book is going to start with a verbal introduction of inverse theory in Chapter 1. Following Chapter 2 the reader can make himself aware of mathematical formulation and properties of inverse problems. By means of a series of linear and nonlinear examples the wide field of inverse problems and applications is outlined. Chapter 3 chooses an optimization approach in order to bridge the gap between identification and control problems formulated in infinite dimensional spaces and associated discretized versions. In Chapter 4, the general discretized inverse problem is solved by Tikhonov's regularization method. Alternative regularization strategies as well as the basic problem of selecting the regularization parameter are also reviewed. Non-Bayesian and Bayesian estimators apply to the stochastic modelling variant in Chapter 5. The final Chapter 6 invites the reader to take some care in evaluating the regularization procedures for nonlinear inverse problems from the point of view of numerical mathematics. Some remarks on computational expense and on software are added. The given list of references includes some important conference proceedings. Moreover, books and papers on inverse theory and applications, but also on auxiliary topics like optimization and statistics, are arranged in alphabetical order of the author's names.

I would like to express my thanks to Professor Dr. V. Friedrich of the Technical University of Karl-Marx-Stadt, who showed me the way to the fascinating field of ill-posed problems ten years ago. Also, I want to acknowledge my linguistic adviser Dr. B. Legler for the valuable corrections and recommendations concerning the English version of this text. Moreover, I feel obliged to Mrs. Dr. R. Müller of Teubner-Verlag for the kind support throughout the elaboration of different stages of the book.

Karl-Marx-Stadt, October 1985 Bernd Hofmann

CONTENTS

## 1. Introduction

Within the last few decades the accelerated coupling of applied mathematics, natural science and engineering has given rise to great interest in computer simulation of real processes. The economic advantage of simulating the behaviour of physical field quantities (e.g. temperature, pressure, stress, velocity, etc.), varying in space and in time, by a digital computer is considerable. Desired properties of a process or desired reactions of a machine can be checked without implementing the expensive process or machinery hardware. Many developments in the fields of numerical mathematics, computer science, analysis, mechanics, system theory etc. have been stimulated by the requirements of practice regarding simulation experiments.

The original simulation problem consists in the numerical determination of effect quantities corresponding to a well-defined causality if all the causal quantities and the relations describing the causality are known. This so-called direct problem needs information about all quantities which influence the unknown effect. Moreover, the inner structure of the causality, all initial and boundary conditions and, finally, the geometric details have to be formulated mathematically. Algebraic and integral formulae, ordinary and partial differential equations or systems of these frequently help to formulate direct problems.

If reality and mathematical formulation coincide sufficiently, then the direct problem is expected to be uniquely solvable. Furthermore, with the exception of some catastrophic processes the effect quantity continuously depends on the cause. Thus a good approximation of the solution to the direct problem will be computable whenever a corresponding mathematical algorithm is implemented and the required input data are precise enough.

Although adequate models and good numerical procedures are used to formulate and solve direct problems, a comparison between real process behaviour and simulation results can indicate that the simulation has failed. This case occurs if the considered causal quantities involve at least one scalar (frequently called parameter) or one function (parameter function) about which information is missing. For example, it does not seem to be useful to compute the direct problem solution with eight significant digits if a parameter of the problem is not even known up to its order of magnitude.

In order to find these unknown values or functions which are not observable directly, one can exploit the causality of the direct problem backwards. Based on some observations (indirect measurements) of the effect quantity, the computer can also help to identify the absent parameters or parameter functions. Problems of this kind are called identification problems. They form the first class of inverse problems. In general, inverse problems aim at determining causal quantities when effect values are given.

Besides identification problems, control problems are also of inverse character. This second class of inverse problems occurs if effect quantities are to be fitted to a desired value, vector or function by controlling the causal quantities over a region of physically or technically admissible elements. Such a control problem appears if the control of initial or boundary values allows us to achieve a desired final state of a system. But in addition to the optimal control of processes, control problems also involve problems of optimal synthesis. They are associated with the design of technical equipment or with the construction of a machine. Here, the causes are formed by the know-how of design and construction, whereas properties of the synthesized object represent the effect. In such a case, the numerical solution of the inverse problem belongs to the class of CAD/CAM problems.

Different causes may yield the same effect. Moreover, causes which are well-separated can result in almost equal effects. This smoothing character of a causality implies a loss of information concerning the inverse problem which is termed ill-posedness. If in the extreme case all causal quantity values yield the same effect, then the loss of information is perfect. For applied inverse problems of identification or control type, the remaining content of information inside the observed effects has to be exploited as far as possible in order to find satisfactory causes. Special methods, the regularization methods, have been developed for this purpose. They will be introduced and characterized in the following chapters of this text in detail.

Already mentioned in papers of TIKHONOV [427] in 1943, KREIN [252] in 1954 and some others, inverse problems really gained ground in mathematics in the 1960's. The first main idea of regularization as a strategy for overcoming the difficulties due to the ill-posedness of inverse problems were outlined independently by TIKHONOV [428] and [429] in 1963 and PHILLIPS [352] in 1962. In the years after TIKHONOV continued to develop and complete this method. At Moscow University he established a mathematical centre dealing with the regularized solution of ill-posed problems. Until now this centre has continuously stimulated the investigations in this field all over the world.

In the 1970's, the monograph of TIKHONOV and ARSENIN [434] (1974 first, 1979 second extended edition) played an important role as a standard work on this topic. Taking into account the requirements of indirect measurement problems, linear integral equations of the first kind and linear operator equations with compact operators were of special interest in this time. In 1973 MOROZOV [312] published a review and bibliography which summarized ten years of mathematical research on regularization especially with respect to Soviet authors.

In 1981 TIKHONOV and MOROZOV [438] gave as a novelty to [312] a report on the recent years (see also the monograph [441]). Today the accelerated development of computer software and hardware allows the successful treatment of identification and control in partial differential equations. The number of papers published on nonlinear inverse problems is permanently growing. They are aimed at the determination of material parameters. In addition, very complex control tasks call for a solution. In this context, solving nonlinear optimization problems becomes a frequently occurring auxiliary problem closely related to the solution of inverse problems. The book of GLASKO [151] in 1984, which is written with respect to physical problems and edited by the Moscow State University, includes many of these aspects.

This introduction will not give an extensive survey of papers on regularization for inverse and ill-posed problems and applications. For the literature we refer to the bibliography at the end of this book, for classes of applied problems, see the series of examples in Chapter 2. However, without striving for completeness we are going to mention briefly some further centres in the Soviet Union, the United States of America and various European countries, from which interesting contributions to the theory and numerical practice of inverse problems have come in recent years.

The Soviet centres for studying inverse problems are Moscow, Novosibirsk (see e.g. LAVRENTIEV [264-8], FEDOTOV [127-8], ISAKOV [219], ROMANOV [370-2]) and Sverdlovsk (see e.g. IVANOV [222-5] and VASIN [461-62]). Mathematicians of universities and academy departments in these cities have successfully dealt with inverse and ill-posed problems for twenty years.

American mathematicians e.g. at the universities of Delaware, Newark (NASHED [327-30], COLTON [72-74]), Ithaca (PAYNE [247]),

Cincinnati (GROETSCH [172-81]), Stanford (GOLUB [159-60]), Madison (WAHBA [466-77]), Baltimore (SEIDMAN [397-99]), Wyoming, Laramie (EWING [121]), Austin (CANNON [49-55]), California, Davis (LINZ [277-79]) and New Brunswick (FALK [123-25]) have substantially contributed to the development of analytical, numerical and statistical methods for these problems. The "Delaware Conference on Ill-Posed Problems", devoted to the theory and practice of solving problems of this kind showed the acceleration of mathematical theory and numerical methods in this topic, but also the wide field of applications in natural sciences (e.g. meteorology, material science, chemistry, geophysics, reservoir stimulation, electrocardiology, control systems design) and engineering.

The Swedish centre at Linköping University (see e.g. BJÖRCK [32-34] and ELDÉN [100-106]) is a residence for ill-posed problems in Europe (see [C5]) just like the University of Linz, Austria (ENGL [108-17]). Since 1975 the treatment of indirect measurements has been of great interest at the Technical University of Karl-Marx-Stadt (see FRIEDRICH [134-39], HOFMANN [198-214] and TAUTENHAHN [421-23]). ANGER's Conference at Halle (GDR) in 1979 (see [C13]) must be mentioned just as the Conference at Oberwolfach (FRG) in 1982 ([C28]) and the Workshop at Heidelberg in 1982 ([C26]). Moreover, the professors LIONS (Paris , [263], [280-81]), SABATIER (Montpellier, [377-78]), CHAVENT (Le Chesnay, [62-66]), MAREK (Prague, [230]), MARTI (Zurich, [293-95]), NATTERER (Munster,[331-40]), HOFFMANN (Augsburg, [193-97]), GORENFLO (Berlin-West, [170]), SCHOCK (Kaiserslautern, [386-91]), DEUFELHARD (Heidelberg,[88]), FRANZONE (Pavia, [132]), DAVIES (Penglais, Wales,[83-84]), KUHNERT (Karl-Marx-Stadt, [255-57]), KLUGE (Berlin, [242-45]), DÜMMEL (Karl-Marx-Stadt,[95-96]), FARZAN (Budapest, [126]), ENGLAND (Mexico-City, [118]), DE HOOG (Canberra, [85-86]), LUKAS (Canberra, [288]), ANDERSSEN (Canberra, [10]) and SUZUKI (Tokyo, [417-19]) have also performed studies on inverse problems during the last ten years. In any year science raises new questions of inverse nature. Therefore, the list of scientists who deal with the mathematical treatment of this problem class will grow considerably in the future.

There are two different routes of solving inverse and ill-posed problems. The common idea of both routes is to replace an ill-posed problem by a well-posed neighbouring problem, the regularization of the inverse problem. In order to obtain a numerical solution by means of a computer, the problem which is generally formulated in mathematical terms by using infinite dimensional function spaces has to be discretized. We say that the continuous route is taken if regularization is performed in infinite dimensional spaces. After that the regularized problem can be discretized. Conversely, "first discretize then regularize" is the characteristic feature of discrete routes. The continuous route as well as the discrete route possess both advantages and disadvantages. This text systematically follows the discrete route in accordance with the intention of the author to handle only real-life data which are indeed available. The question of how to discretize an inverse problem in an optimal manner cannot be answered by this text. But provided a discretized inverse problem is given, many details which promote the numerical construction of approximate solutions will be described, motivated and characterized in terms of numerical analysis here.

## 2. Mathematical Modelling

### 2.1. Formulation, Discretization and Properties of Inverse Problems

We are now going to describe inverse problems in a general form. In the sequel, identification problems that form the first class of inverse problems will be expressed by an operator equation

$$\mathcal{A}\varkappa = b\ , \varkappa \in \mathcal{D} \subseteq B_1\ ,\ b \in B_2\ . \tag{2.1}$$

Assumption 2.1:
Throughout this book let $B_1$ and $B_2$ be real Banach spaces with norms $\|\cdot\|_1$ and $\|\cdot\|_2$, respectively. Furthermore, we assume $\mathcal{A}: \mathcal{D} \subseteq B_1 \to B_2$ to be a continuous operator defined on a closed convex set $\mathcal{D}$ of admissible solutions. ○

Definition 2.2:
The problem (2.1) is said to be a linear identification problem and can be considered as the first special case of a linear inverse problem if $\mathcal{A}$ is linear in the following sense:

$$\begin{gathered}\mathcal{A}\varkappa^{(3)} = \lambda_1 \cdot \mathcal{A}\varkappa^{(1)} + \lambda_2 \cdot \mathcal{A}\varkappa^{(2)} \quad \text{if} \quad \varkappa^{(3)} = \lambda_1 \cdot \varkappa^{(1)} + \lambda_2 \cdot \varkappa^{(2)}, \\ \lambda_1, \lambda_2 \in R,\ \varkappa^{(1)}, \varkappa^{(2)}, \varkappa^{(3)} \in \mathcal{D}\ .\end{gathered} \tag{2.2}$$

If $\mathcal{A}$ is nonlinear, then (2.1) is a nonlinear identification problem. We call $\mathcal{A}$ the operator of the direct problem and distinguish unconstrained and constrained identification problems according to $\mathcal{D} = B_1$ and $\mathcal{D} \neq B_1$, respectively. ○

The elements $\varkappa$ represent the unknown physical quantities which are to be determined from information about the observable quantity $b$ in the right-hand side of Eq. (2.1). Transformations $\mathcal{A}$ may be given explicitly, e.g. by an integral operator possessing a known kernel. However, they can also be defined implicitly. This implicit form occurs if parameter functions are to be identified in partial differential equations. Then $\mathcal{A}\varkappa$ is the solution to a boundary or initial-boundary value problem, e.g. of heat equation, and $\varkappa$ a corresponding parameter in the differential equation, in the initial condition or in the boundary condition. We say that the direct problem is solved if the element $\mathcal{A}\varkappa$ is calculated for a given element $\varkappa \in \mathcal{D}$.

Sometimes control problems can be written in the form (2.1) and interpreted as identification problems. However, a more general notation is the optimization formulation:

$$\underset{\varkappa \in \mathcal{D}}{\text{minimize}} \|\mathcal{A}\varkappa - b\|_2\ . \tag{2.3}$$

Here, the element $\varkappa$ is the control element, which we have to find so that a desired effect quantity $\ell$ is approximated as best as possible. The measure of distance between the obtained and desired effect magnitudes is assumed to be the norm of the space $B_2$.

The nature of inverse problems (see Chapter 1) leads to ill-posedness as a characteristic property of these problems. We have to know this property in order to overcome arising difficulties during the solution process. Therefore we formulate the well-posedness and ill-posedness definition according to HADAMARD [184] for identification problems (2.1). We go back to VASILEV [460] for control problems (2.3). However beforehand, we will introduce some notation in order to avoid misunderstanding in the following parts of this book.

Definition 2.3:
We define linear, nonlinear, unconstrained and constrained control problems analogously to Def. 2.2 by the linearity or nonlinearity of $\mathcal{A}$ and with respect to the constraints given by $\mathcal{D}$. ○

Definition 2.4:
An operator $\mathcal{A}: \mathcal{D} \subseteq B_1 \rightarrow B_2$ is injective over $\mathcal{D}$ if $\mathcal{A}\varkappa^{(1)} = \mathcal{A}\varkappa^{(2)}$ with $\varkappa^{(1)}, \varkappa^{(2)} \in \mathcal{D}$ implies the equation $\varkappa^{(1)} = \varkappa^{(2)}$. It is called surjective over $\mathcal{D}$ if $\mathcal{A}\mathcal{D} = B_2$, and bijective over $\mathcal{D}$ if this operator is injective and surjective over $\mathcal{D}$. For an injective operator $\mathcal{A}$ over $\mathcal{D}$ we have a uni-valued mapping $\mathcal{A}^{-1}: \mathcal{A}\mathcal{D} \rightarrow \mathcal{D}$, which forms the inverse operator to $\mathcal{A}$. If $\mathcal{A}$ fails to be injective, $\mathcal{A}^{(-1)}$ will indicate the multi-valued mapping $\ell \longrightarrow \mathcal{A}^{(-1)}\ell$ transforming an element $\ell \in \mathcal{A}\mathcal{D}$ into the full inverse image set $\mathcal{A}^{(-1)}\ell$. Finally, $\mathcal{X}(\ell)$ will denote the set of all solutions to a control problem (2.3) for given $\ell$. In the case of linear operators, instead of $\mathcal{A}^{-1}$ we use the symbol $\mathcal{A}_L^{-1}$ in order to point out that we consider a left inverse with $\mathcal{A}_L^{-1} \cdot \mathcal{A}_L = I$ (I-unity operator). In addition, $\mathcal{A}^+: B_2 \rightarrow B_1$ denotes the pseudoinverse operator of $\mathcal{A}$ (see GROETSCH [177]). In this linear case $\mathcal{A}^{-1}$ is used as the symbol of a left and right inverse operator $\mathcal{A}^{-1}\mathcal{A} = \mathcal{A}\mathcal{A}^{-1} = I$ of a bijective operator $\mathcal{A}: B_1 \rightarrow B_1$. The symbols $\mathbb{N}(\mathcal{A}) = \{\varkappa \in B_1 : \mathcal{A}\varkappa = 0\}$ for the null-space and $\mathbb{R}(\mathcal{A}) = \{\ell \in B_2 : \ell = \mathcal{A}\varkappa, \varkappa \in B_1\}$ for the range are needed if unconstrained linear inverse problems occur. If we consider operators in finite dimensional spaces, then the symbols introduced above are also used. ○

Now we take up the well-posedness property for identification and control problems:

Definition 2.5:

The problem (2.1) is called well-posed if the following three conditions hold:

i. $\mathcal{A}$ is surjective over $\mathcal{D}$ (existence condition).

ii. $\mathcal{A}$ is injective over $\mathcal{D}$ (uniqueness condition).

iii. Given $b \in B_2$ and $\varepsilon > 0$, we can choose $\delta = \delta(\varepsilon, b) > 0$ so that $\|x^{(1)} - x\|_1 < \varepsilon$ whenever $\|\mathcal{A}x^{(1)} - b\|_2 < \delta$, $\mathcal{A}x = b$ and $x, x^{(1)} \in \mathcal{D}$ (stability condition).

If at least one of these conditions is injured, then (2.1) is said to be ill-posed. ○

Definition 2.6:

The problem (2.3) is called well-posed if the following conditions are both satisfied:

iv. The set $\mathcal{X}(b) := \left\{ x^{(1)} \in \mathcal{D} : \|\mathcal{A}x^{(1)} - b\|_2 = \inf_{x \in \mathcal{D}} \|\mathcal{A}x - b\|_2 \right\}$ is nonempty for all $b \in B_2$ (existence condition).

v. Given $b \in B_2$ and $\varepsilon > 0$, we can choose $\delta = \delta(\varepsilon, b) > 0$ so that $\operatorname{dist}(x, \mathcal{X}(b)) := \inf_{x^{(1)} \in \mathcal{X}(b)} \|x - x^{(1)}\|_1 < \varepsilon$ whenever $\|\mathcal{A}x - b\|_2 < \inf_{x^{(1)} \in \mathcal{D}} \|\mathcal{A}x^{(1)} - b\|_2 + \delta$ and $x \in \mathcal{D}$ (stability condition).

Otherwise, the problem is said to be ill-posed. ○

Remark 2.7:

The stability condition iii. is valid if and only if $\mathcal{A}$ is injective over $\mathcal{D}$ and has a continuous inverse. On the other hand, the stability condition v. does not require an injective operator $\mathcal{A}$. This kind of stability corresponds to the continuity of the multi-valued inverse mapping of $\mathcal{A}$. The uniqueness of the solution to an identification problem (2.1) is desirable, since any physical quantity that is to be found is uniquely determined. Thus, injectivity of $\mathcal{A}$, for identification problems, indicates that mathematical modelling is sufficiently adequate. Conversely, various control elements can yield the same effect. In this case, noninjective operators $\mathcal{A}$ are natural for problem (2.3). The well-posedness definitions given above take into account this difference. ○

In this text, we are especially interested in handling real life data by a digital computer. Therefore, we essentially consider the situation that only a finite number m of data is available as information about $b$. In consequence of this, we mostly acquiesce in determining a skeleton solution of $\mathscr{X}$ based on n values. This finitization will be formulated by establishing a pair of projection operators $P^n$ for a priori discretization and $Q^m$ for the experimental design (e.g. sensor location spatially and in time).

Assumption 2.8:
In the sequel, let $P^n: B_1 \to R^n$ and $Q^m: B_2 \to R^m$ be a pair of bounded linear operators. We denote by $x=(x_1,x_2,\ldots,x_n)^T =: P^n \mathscr{X} \in R^n$ and by $b=(b_1,b_2,\ldots,b_m)^T =: Q^m \mathscr{B} \in R^m$ the image vectors of these projections. Furthermore, the set $D := P^n \mathscr{D}$ describes the set of admissible solutions in the n-dimensional space. Note that any of the components $x_i$, $i=1(1)n$ and $b_j$, $j=1(1)m$ represents a linear bounded functional of $\mathscr{X}$ and $\mathscr{B}$, respectively. By Assumption 2.1, D is a closed and convex subset of $R^n$. o

In order to evaluate the quality of solutions obtained, we also introduce a back projection operator $\mathscr{P}^n: D \subseteq R^n \longrightarrow \mathscr{D} \subseteq B_1$ that transformes any skeleton solution vector $x \in D$ into an associated Banach space element $\mathscr{X} \in \mathscr{D}$.

Provided that a finitization as above is used to deal with inverse problems, any strategy of fitting $x \in D$ by means of b requires an operator $A: D \subseteq R^n \to R^m$ connecting both vectors. In such a way,

$$A\,x = b\,, \quad x \in D \subseteq R^n, \quad b \in R^m \tag{2.4}$$

represents a discrete version of problem (2.1). A corresponding version of problem (2.3) is

$$\underset{x \in D}{\text{minimize}}\ \| A\,x - b \|^2 \,. \tag{2.5}$$

Since the treatment of expressions of the form $\| A\,x - b \|^2$ is easier than the treatment of $\| A\,x - b \|$ , for simplicity we will use the Euclidean residual norm square instead of the residual norm itself.

Definition 2.9:
We call the problems (2.4) and (2.5) discretized inverse problems (discretized identification and control problems, respectively). Throughout this text, $\| . \|$ stands for the Euclidean norm, $( .,. )$ for the associated inner product in a finite dimensional space. If misunderstandings are excluded, we do not mark the dimension number as a right-hand corner subscript of the norm symbol. o

As far as possible we focus our attention to Euclidean norms and least squares solutions to discretized inverse problems. A generalization of lemmas and theorems derived in this text to other norms can often be found easily.

Assumption 2.10:

In the sequel, let A: $D\subseteq R^n \to R^m$ be a continuous operator chosen so that the composite operator $AP^n: \mathcal{D}\subseteq B_1\to R^m$ approximates the operator $Q^m\mathcal{A}: \mathcal{D}\subseteq B_1 \to R^m$. ○

Remark 2.11:

The discretization of $\varkappa$ and $\mathcal{A}$, for given $Q^m$, is the better the smaller the associated deviations $\|AP^n\varkappa - Q^m\mathcal{A}\varkappa\|$ are. Given $\mathcal{A}$, $P^n$ and $Q^m$, the choice of A should try to carry over intrinsic properties (e.g. linearity or monotonicity) from $\mathcal{A}$ to A. ○

Definition 2.12:

We define linear, nonlinear, constrained and unconstrained discretized inverse problems analogously to Definition 2.2. If we replace the symbols $\mathcal{A}$, $\mathcal{D}$, $\varkappa$, $\mathcal{b}$, $\|\cdot\|_1$ and $\|\cdot\|_2$ by A, D, x, b and $\|\cdot\|$, respectively, the Definitions 2.5 and 2.6 apply to discretized inverse problems in order to distinguish well-posed and ill-posed discretized identification and control problems. ○

The injectivity and surjectivity of an operator $\mathcal{A}$ does not imply the same property for the discrete version A. Existence and uniqueness of a solution to problems (2.4) depend upon the concrete form of A and essentially upon the dimension numbers m and n and their proportion. If n gets large enough and the continuous problem (2.1) is ill-posed so that the stability condition iii. is not valid, then (2.4) tends to become ill-conditioned. That means, small perturbations of the input data b may lead to large changes in the solution x. Therefore, substantial difficulties arise in solving these problems numerically in view of the fact that any observation generates errors and thus the input data are always noisy.

We present a variant of RHEINBOLDT's definition [364] of the numerical condition to problem (2.4).

Definition 2.13:

A problem (2.4) is called well-conditioned at a solution $x\in D$ satisfying the equation $Ax = b$ if the condition number

$$\operatorname{cond}_1(A,S,x) := \sup_{x\neq x^{(1)}\in S} \frac{\|Ax^{(1)}-b\|}{\|x^{(1)}-x\|} \Bigg/ \inf_{x\neq x^{(2)}\in S} \frac{\|Ax^{(2)}-b\|}{\|x^{(2)}-x\|} \tag{2.6}$$

is small for a sufficiently small closed sphere $S=\bar{S}(x,\varepsilon)\subseteq R^n$ with centre x and radius $\varepsilon>0$, otherwise it is termed ill-conditioned at x. ○

Lemma 2.14:

If (2.4) is a linear unconstrained discretized inverse problem with rank(A)=n, then

$$\mathrm{cond}_1(A,S,x) = \mathrm{cond}_1(A) := \|A\|\|A^+\| = \frac{\sigma_1}{\sigma_n} \tag{2.7}$$

is independent of x and $\varepsilon$. The symbols $\|A\|$ and $\sigma_i$ refer to the spectral norm and to the i-th largest singular value of the matrix A, respectively. $A^+$ denotes the Moore-Penrose inverse to A (for properties see KUHNERT [255] and ZIELKE [489] ). ○

Proof:

In the sequel, we identify the operator A and its matrix representation. Let $\{\sigma_i, u^{(i)}, v^{(i)}\}$ , i=1,2,...,r=rank(A) denote a singular system of A consisting of singular values $\sigma_1 \geq \sigma_2 \geq \ldots \geq \sigma_r > 0$, which are arranged in order of magnitude, and associated pairs of singular vectors that can be completed to a pair of complete orthonormal systems $\{u^{(i)}\}_{i=1}^n \subseteq R^n$ and $\{v^{(i)}\}_{i=1}^m \subseteq R^m$ (see e.g. BJÖRCK [32] ). Using the symbol $A^T$ for matrix transposition, we obtain

$$\left.\begin{array}{l} A\cdot u^{(i)} = \sigma_i \cdot v^{(i)}, \quad A^T\cdot v^{(i)} = \sigma_i\cdot u^{(i)}, \qquad 1\leq i\leq r; \\ A\cdot u^{(i)} = 0, \quad r+1\leq i\leq n; \quad A^T\cdot v^{(i)} = 0, \quad r+1\leq i\leq m; \\ A^+\cdot v^{(i)} = \dfrac{1}{\sigma_i}\cdot u^{(i)}, \quad 1\leq i\leq r; \quad A^+\cdot v^{(i)} = 0, \quad r+1\leq i\leq m; \\ \|A\| = \sigma_1, \quad \|A^+\| = \dfrac{1}{\sigma_r}. \end{array}\right\} \tag{2.8}$$

In this proof, we have to consider the case r=n. Then, in view of the fact that any sphere with centre x contains vectors $x^{(1)}$, $x^{(2)}$ with $\dfrac{x^{(1)}-x}{\|x^{(1)}-x\|} = u^{(1)}$ and $\dfrac{x^{(2)}-x}{\|x^{(2)}-x\|} = u^{(n)}$, we have

$$\sup_{x\neq x^{(1)}\in S} \frac{\|Ax^{(1)}-Ax\|}{\|x^{(1)}-x\|} = \sup_{0\neq x'\in R^n} \frac{\|Ax'\|}{\|x'\|} = \|A\| = \sigma_1 \quad \text{and} \quad \sup_{0\neq x'\in R^n} \frac{\|A^+Ax'\|}{\|Ax'\|} =$$

$$\left(\inf_{x\neq x^{(2)}\in S} \frac{\|Ax^{(2)}-Ax\|}{\|x^{(2)}-x\|}\right)^{-1} = \|A^+\| = \sigma_n^{-1}. \text{ Therefore, } \mathrm{cond}_1(A,S,x) =$$

$= \|A\|\|A^+\| = \dfrac{\sigma_1}{\sigma_n}$. To explain the above equations, we note that due to rank(A)=n , for all $x'\in R^n$, $x' = A^+A\,x'$. #

Remark 2.15:

Only a small class of discretized identification problems satisfies the existence condition i. Namely, $AD = R^m$ is not typical for nonlinear operators A. Linear operators A are surjective over $R^n$ if and only if rank(A)=m (nonoverdetermined problems). On the other hand, condition ii. is a desired property of discretized identification problems. Linear unconstrained problems possess this property iff we have $\mathbb{N}(A) = \{0\}$ (full-rank problems). Finally, note that there is no linear discretized identification problem where A is injective and $A_L^{-1}: \mathbb{R}(A) \to R^n$ is a discontinuous (unbounded) linear mapping. This is a consequence of the fact that no unbounded linear operators exist with domain and range in finite dimensional spaces. Therefore, condition ii. implies condition iii. in the linear finite dimensional case. However, there are nonlinear injective operators A and sets D such that (2.4) is instable. Then, this ill-posedness is extreme situation of a sequence of ill-conditioned problems the condition numbers of which tend to infinity. The inverse value of the denominator of $cond_1(.)$ is a measure of the continuity of $A^{-1}$. If this measure grows without bound, then the problem becomes instable. o

Now we will investigate the numerical condition of control problems (2.5). Note that the existence condition iv. holds if the set AD is closed. Under the assumptions here stated, for linear operators A, this is always true. Namely, the linear transformation of a closed set of $R^n$ into a Banach space always possesses a closed range in this space (cf. [249, Sec. 3.3.2]).

The symbol X(b) will refer to the set of solutions to problem (2.5) for given $b \in R^m$. We introduce the quasidistance of two sets X and Y as follows:

$$\operatorname{qdist}(X,Y) := \sup_{x \in X} \operatorname{dist}(x,Y) \tag{2.9}$$

(for properties see IVANOV [222] ). Let us exclude the pathological case $Ax \equiv \text{const.} \in R^m$ for all $x \in D$.

Definition 2.16:

A problem (2.5) is called well-conditioned at $b \in R^m$ if the generalized condition number

$$\operatorname{cond}_2(A,T,b) := = \sup_{b^{(1)} \in T_b} \frac{\operatorname{qdist}(AX(b), AX(b^{(1)}))}{\operatorname{qdist}(X(b), X(b^{(1)}))} \Bigg/ \inf_{b^{(2)} \in T_b} \frac{\| b - b^{(2)} \|}{\operatorname{qdist}(X(b), X(b^{(2)}))} \tag{2.10}$$

with $T_b=\{b'\in T: X(b')\neq X(b)\}$ is small for a sufficiently small closed sphere $T=\bar{T}(b,\varepsilon)\subseteq R^m$ with centre b and radius $\varepsilon>0$. Otherwise, the problem is called ill-conditioned at b. ○

Lemma 2.17:

If (2.5) is a linear unconstrained discretized control problem, then we have, for r=rank(A),

$$\mathrm{cond}_2(A,T,b)=\mathrm{cond}_2(A):=\|A\|\cdot\|A^+\|=\frac{\sigma_1}{\sigma_r} \tag{2.11}$$

with a condition number independent of b and $\varepsilon$. ○

Proof:

We use the singular value decomposition of the proof of Lemma 2.14. Here, the noninjective case $r<n$ may occur. Since the case $A=0$ has been excluded, consider $r>0$. We obtain $X(b)=\{x\in R^n: x=A^+b+x', x'\in \mathbb{N}(A)\}$ due to $\mathbb{N}(A)\oplus \mathbb{R}(A^+)=R^n$ and $\|A^+b+x'-(A^+b^{(1)}+x'')\|^2=$ $=\|A^+(b-b^{(1)})\|^2+\|x'-x''\|^2$ with $x', x''\in\mathbb{N}(A)$. Thus, $\mathrm{qdist}(X(b),X(b^{(1)}))=\|A^+(b-b^{(1)})\|$ and $\mathrm{qdist}(AX(b),AX(b^{(1)}))=$ $=\|AA^+(b-b^{(1)})\|$. By using the expansions $w=\sum_{i=1}^{m}(w,v^{(i)})\cdot v^{(i)}$, $A^+w=\sum_{i=1}^{r}\frac{1}{\sigma_i}\cdot(w,v^{(i)})\cdot u^{(i)}$ and $AA^+w=\sum_{i=1}^{r}(w,v^{(i)})\cdot v^{(i)}$, for an arbitrarily chosen vector $w\in R^m$, it is easy to derive that $\mathrm{cond}_2(A)=$ $=\|A\|\|A^+\|=\frac{\sigma_1}{\sigma_r}$. In this context, one has to consider the relation $T_b=\{b'\in T: A^+(b'-b)\notin\mathbb{N}(A)\}=T\cap(R^m\setminus\{b'\in R^m: A^+b'=A^+b\})$ and the existence of a pair of vectors $b^{(1)}$ and $b^{(2)}$ satisfying the equations $\frac{b^{(1)}-b}{\|b^{(1)}-b\|}=v^{(1)}$ and $\frac{b^{(2)}-b}{\|b^{(2)}-b\|}=v^{(r)}$ in any sphere with centre b.#

Theorem 2.18:

Any linear unconstrained discretized control problem (2.5) is well-posed. ○

Proof:

Due to $X(b)=\{x\in R^n: x=A^+b+x_N, x_N\in\mathbb{N}(A)\}\neq\emptyset$, the existence condition iv. holds. We still have to prove the stability condition v. Assume that there were a sequence of vectors $\{x^{(i)}\}_{i=1}^{\infty}\subseteq R^n$ satisfying the inequalities $\mathrm{dist}(x^{(i)},X(b))\geq\varepsilon$, for all i, and the limiting condition $\lim_{i\to\infty}\|Ax^{(i)}-\tilde{b}\|=0$, $\tilde{b}:=AA^+b$. In view of $R^n=\mathbb{R}(A^+)\oplus\mathbb{N}(A)$, for

$b^{(i)}=Ax^{(i)}$ and all i, we would have a representation $x^{(i)}=A^+b^{(i)}+x_N^{(i)}$ with $x_N^{(i)}\in \mathbb{N}(A)$. This would also imply $\mathrm{dist}(x^{(i)},X(b))=\|A^+(b^{(i)}-\tilde{b})\|\geq\varepsilon$ and hence $\inf_i \|b^{(i)}-\tilde{b}\|>0$, since the inequality

$$\sup_i \frac{\|A^+(b^{(i)}-\tilde{b})\|}{\|b^{(i)}-\tilde{b}\|}\leq\|A^+\|<\infty$$

is valid. Thus, we obtain a contradiction. Consequently, the problem (2.5) is always well-posed if A is linear and $D=R^n$ holds. #

Remark 2.19:

Note that linear constrained discretized control problems (2.5) may be ill-posed if the set D is unbounded and fails to be a linear manifold. Examples for nonlinear discretized inverse problems of control type are given in [460]. The inverse value of the denominator of $\mathrm{cond}_2(.)$ is a stability measure of the multi-valued mapping $b\rightarrow X(b)$. If this denominator tends to zero, the problems become more and more ill-conditioned, where $\mathrm{cond}_2(.)=\infty$ represents the ill-posed extreme situation. Furthermore, note that, in consequence of Theorem 2.18, a linear control problem (2.3) with $\mathfrak{D}=\mathrm{span}(e^{(1)},\ldots,e^{(n)})$, $e^{(i)}\in B_1$, $i=1(1)n$, is always well-posed. o

Theorem 2.20:

Let $A: D\subseteq R^n\rightarrow R^m$ be injective and, for a sphere S with centre x, $\mathrm{cond}_1(A,S,x)<\infty$. Then

$$\frac{\|x-x^{(1)}\|}{\|x-x^{(0)}\|}\leq \mathrm{cond}_1(A,S,x)\cdot\frac{\|Ax-Ax^{(1)}\|}{\|Ax-Ax^{(0)}\|} \tag{2.12}$$

if $x^{(0)}, x^{(1)}\in S$ and $x, x^{(0)}, x^{(1)}$ distinct. o

Proof:

We get $\|x-x^{(1)}\|\leq \sup\limits_{x\neq x^{(2)}\in S}\dfrac{\|x-x^{(2)}\|}{\|Ax-Ax^{(2)}\|}\cdot\|Ax-Ax^{(1)}\|$ and

$$\frac{\|Ax-Ax^{(0)}\|}{\|x-x^{(0)}\|}\leq\sup_{x\neq x^{(3)}\in S}\frac{\|Ax-Ax^{(3)}\|}{\|x-x^{(3)}\|}$$

. If the above two inequalities are multiplied, then an inequality equivalent to (2.12) appears. #

Thus, the condition number can be interpreted as a multiplier connecting the relativized errors of the right-hand side (input data) and of the solution with respect to an origin vector $x^{(0)}$. Whereas, for nonlinear problems, the inequality (2.12) is only valid in a neighbourhood of a point x, the global character of $\mathrm{cond}_1(A)$ in the linear case allows an inequality of the form

$$\frac{\|x-x'\|}{\|x\|} \leq \mathrm{cond}_1(A) \cdot \frac{\|Ax-Ax'\|}{\|Ax\|} = \|A\|\|A^+\| \cdot \frac{\|Ax-Ax'\|}{\|Ax\|}, \quad x \notin \mathbb{N}(A). \tag{2.13}$$

A similar relation holds when we are concerned with problem (2.5).

<u>Theorem 2.21:</u>
Let be $T \subseteq R^m$ with centre b and $\mathrm{cond}_2(A,T,b) < \infty$. Then

$$\frac{\mathrm{qdist}(X(b),X(b^{(1)}))}{\mathrm{qdist}(X(b),X(b^{(0)}))} \leq \mathrm{cond}_2(A,T,b) \cdot \frac{\|b-b^{(1)}\|}{\mathrm{qdist}(AX(b),AX(b^{(0)}))} \tag{2.14}$$

if $b^{(0)}, b^{(1)} \in T_b$ (see Definition 2.16). ○

<u>Proof:</u>
We obtain

$$\mathrm{qdist}(X(b),X(b^{(1)})) \leq \sup_{b^{(2)} \in T_b} \frac{\mathrm{qdist}(X(b),X(b^{(2)}))}{\|b-b^{(2)}\|} \cdot \|b-b^{(1)}\| \quad \text{and}$$

$$\frac{\mathrm{qdist}(AX(b),AX(b^{(0)}))}{\mathrm{qdist}(X(b),X(b^{(0)}))} \leq \sup_{b^{(3)} \in T_b} \frac{\mathrm{qdist}(AX(b),AX(b^{(3)}))}{\mathrm{qdist}(X(b),X(b^{(3)}))}.$$ Again, the

multiplication of both inequalities yields the estimate desired in formula (2.14).#

The special case of linear control problems allows a generalization of the local inequality (2.14) with origin element $b^{(0)}=0$ to a global inequality

$$\frac{\mathrm{qdist}(X(b),X(b^{(1)}))}{\|A^+b\|} \leq \mathrm{cond}_2(A) \cdot \frac{\|b-b^{(1)}\|}{\|AA^+b\|}, \quad b \notin \mathbb{N}(A^+), \tag{2.15}$$

where $\|A^+b\| = \min_{x \in X(b)} \|x\|$, $\|AA^+b\| \equiv \|Ax\|$ for each $x \in X(b)$.

Therefore, independently of the concrete type of discretized inverse problem, ill-conditioning may lead to large changes in the solution when the input data are perturbed.

## 2.2. A Survey of the Ill-Posedness Phenomenon via Examples

### 2.2.1. Linear Fredholm Integral Equations of the First Kind in Banach Spaces

An important class of linear inverse problems (2.1) is described by a Fredholm integral equation of the first kind

$$\int_{\underline{t}}^{\bar{t}} k(s,t)\cdot x(t)\cdot dt = b(s), \quad \underline{s}\le s\le\bar{s}, \quad \underline{t}\le t\le\bar{t}, \quad k(s,t)\ge\underline{k}>0, \tag{2.16}$$

where $k(s,t)$ is a given smooth kernel function (at least continuous with respect to $s$ and $t$). Let $\mathfrak{x} = \langle x(t), \underline{t}\le t\le\bar{t}\rangle \in C[\underline{t},\bar{t}] =: B_1$ and $\mathfrak{b} = \langle b(s), \underline{s}\le s\le\bar{s}\rangle \in C[\underline{s},\bar{s}] =: B_2$ be continuous functions and let $\mathcal{A}: \mathcal{D} \subseteq B_1 \to B_2$ be the integral operator according to (2.16) with domain $\mathcal{D} := \{ \mathfrak{x}\in C[\underline{t},\bar{t}]: x(t)\ge 0, \underline{t}\le t\le\bar{t} \}$.

Problems of this form, for example, occur in remote sounding, determination of vertical temperature profiles (see Sec. 5.3.), in geophysics, optics and nuclear physics (see e.g. LAVRENTEV [264], DEUFELHARD/HAIRER [C26], HÄMMERLIN/HOFFMANN [C28] and HERMANN/NATTERER [C19], also [98], [201], [250], [453] and [C12]). An unknown nonnegative function $\mathfrak{x}$ (e.g. temperature, number of particles) is to be identified from indirect measurements of another function $\mathfrak{b}$ (e.g. radiation intensity, number of pulses). A law of nature connecting $\mathfrak{x}$ and $\mathfrak{b}$ forms the kernel function $k(s,t)$ (e.g. transmissitivity of transport media).

In this section, we are going to discretize the problem by numerical quadrature based on the rectangular rule and by collocation. Let

$$\begin{aligned} t_i &:= \underline{t} + \left(\frac{2i-1}{2}\right)\cdot h, \quad h := \frac{\bar{t}-\underline{t}}{n}, \quad i=1(1)n, \\ s_j &:= \underline{s} + j\cdot\tau, \quad \tau := \frac{\bar{s}-\underline{s}}{m}, \quad j=1(1)m \end{aligned} \tag{2.17}$$

be a pair of equally spaced grids for the intervals $[\underline{t},\bar{t}]$ and $[\underline{s},\bar{s}]$. We define

$$\begin{aligned} P^n\mathfrak{x} &:= x = (x(t_1), x(t_2), \ldots, x(t_n))^T, \\ Q^m\mathfrak{b} &:= b = (b(s_1), b(s_2), \ldots, b(s_m))^T \end{aligned} \tag{2.18}$$

and

$$(Ax)_j := \sum_{i=1}^{n} h\cdot k(s_j,t_i)\cdot x_i, \quad j=1(1)m, \tag{2.19}$$

where the linear operator A is expressed by its matrix representation $A := (a_{ji})_{\substack{j=1(1)m\\ i=1(1)n}}$, $a_{ji} := h\cdot k(s_j,t_i)$. The linearity of $\mathcal{A}$ carries over to A.

As the back projection operator $\wp^n$ we can use

$$(\wp^n x)(t) := \sum_{i=1}^{n} x_i\cdot \varphi^{(i)}(t) \tag{2.20}$$

with basis functions $\varphi^{(i)}(t) \in C[\underline{t},\bar{t}]$, $i=1(1)n$. The error of discretization $\|Q^m\mathcal{A}\varkappa - AP^n\varkappa\|$ tends to zero as $n\to\infty$. Nonnegativity constraints, which form $\mathcal{D}$, imply again nonnegativity constraints $D:= P^n\mathcal{D} = \{x\in R^n: x_i \geq 0,\ i=1(1)n\}$ in the discrete version. Now we consider some properties of Eq. (2.16) and of its discrete analgoue (2.4) according to (2.17) through (2.19).

Definition 2.22:

A continuous operator $\mathcal{A}:\mathcal{D}\subseteq B_1\to B_2$ is said to be compact iff, for any bounded subset $\mathcal{M}\subseteq\mathcal{D}$, the range $\mathcal{A}\mathcal{M}$ is a relatively compact subset of $B_2$. o

Remark 2.23:

The integral operator $\mathcal{A}$ of the problem (2.16) is a linear bounded (i.e., continuous) and compact operator from $B_1$ into $B_2$ (see KANTOROVICH/AKILOV [236, p.245]) with

$$\|\mathcal{A}\|:= \sup_{0\neq\varkappa}\frac{\|\mathcal{A}\varkappa\|_2}{\|\varkappa\|_1} = \max_{\underline{s}\leq s\leq\bar{s}}\int_{\underline{t}}^{\bar{t}} |k(s,t)|\,dt \quad ([236,\ p.90]).$$

We will show that the problem (2.16) subject to nonnegativity constraints is an ill-posed identification problem. o

Lemma 2.24:

If $B_1$ is an infinite dimensional Banach space, then there is a bounded subset of $B_1$, which fails to be relatively compact. o

Proof: LJUSTERNIK/SOBOLEV [284, p.172].

Lemma 2.25:

Let (2.1) be a linear unconstrained problem with a bounded compact and injective operator $\mathcal{A}: B_1\to B_2$. Moreover, let $B_1$ be infinite dimensional. Then $\mathcal{A}_L^{-1} : \mathcal{R}(\mathcal{A})\subseteq B_2\to B_1$ is a linear unbounded (i.e., discontinuous) operator. o

Proof:

Assume that the obviously linear operator $\mathcal{A}_L^{-1}$ is unbounded, i.e.,

$$\|\mathcal{A}_L^{-1}\| := \sup_{0\neq \mathscr{b}\in\mathcal{R}(\mathcal{A})}\frac{\|\mathcal{A}_L^{-1}\mathscr{b}\|_1}{\|\mathscr{b}\|_2} = \sup_{0\neq\varkappa\in B_1}\frac{\|\varkappa\|_1}{\|\mathcal{A}\varkappa\|_2} < \infty$$

. Owing to Lemma 2.24 we can choose a bounded sequence $\{\varkappa^{(i)}\}_{i=1}^{\infty}\subseteq B_1$ that fails to be relatively compact, however the associated sequence of range elements $\{\mathscr{b}^{(i)}\}_{i=1}^{\infty} = \{\mathcal{A}\varkappa^{(i)}\}_{i=1}^{\infty}\subseteq B_2$ has a point of accumulation in $B_2$. Without loss of generality we assume that $\lim_{i\to\infty}\mathscr{b}^{(i)} = \mathscr{b}\in B_2$. Then, $\{\mathscr{b}^{(i)}\}_{i=1}^{\infty}$ is a Cauchy sequence, i.e., given $\varepsilon$ and $\delta = \frac{\varepsilon}{\|\mathcal{A}_L^{-1}\|}$, there is an integer $i_0$ so that $\|\mathscr{b}^{(i)} - \mathscr{b}^{(j)}\|_2 \leq \delta$ if $i,j\geq i_0$.

Moreover, $\|\mathscr{x}^{(i)} - \mathscr{x}^{(j)}\|_1 \leq \varepsilon$ and therefore $\{\mathscr{x}^{(i)}\}_{i=1}^{\infty}$ also proved to be a Cauchy sequence. This is a contradiction. #

A problem (2.1) satisfying the assumptions of Lemma 2.25 is ill-posed since the stability condition iii. cannot hold whenever $\|\mathcal{A}_L^{-1}\| = \infty$. For the unconstrained fashion of this problem (2.16), the instability can also be shown by the Riemann lemma

$\int_{\underline{t}}^{\bar{t}} k(s,t)\cdot \sin \lambda t \cdot dt \to 0$ as $\lambda \to \infty$, where small right-hand side changes of this form correspond to arbitrarily large solution changes as $\lambda \to \infty$. However, it may occur that problems become stable in the constrained case. For example, this is evident for a compact domain $\mathcal{D}$ as the following so-called Tikhonov theorem indicates.

Theorem 2.26:

If $\mathcal{A}: \mathcal{D} \subseteq B_1 \to B_2$ is a continuous operator on a compact domain $\mathcal{D} \subseteq B_1$, then the multi-valued mapping $\mathcal{A}^{(-1)}: \mathcal{A}\mathcal{D} \to 2^{\mathcal{D}}$ ($2^{\mathcal{D}}$ - power set of $\mathcal{D}$) is continuous in the following sense:

$$\lim_{i\to\infty} \operatorname{qdist}\,(\mathcal{A}^{(-1)}\, \mathscr{b}^{(i)}, \mathcal{A}^{(-1)}\mathscr{b}) = 0 \qquad (2.21)$$

if $\lim_{i\to\infty} \|\mathscr{b}^{(i)} - \mathscr{b}\|_2 = 0$, $\mathscr{b}, \mathscr{b}^{(i)} \in \mathcal{A}\mathcal{D}$, $i=1,2,\ldots$. That means, the uni-valued inverse mapping $\mathcal{A}^{-1}: \mathcal{A}\mathcal{D} \to \mathcal{D}$ is a continuous operator if $\mathcal{A}$ is injective over $\mathcal{D}$. o

Proof:

Consider an arbitrarily chosen sequence of elements $\mathscr{x}^{(i)} \in \mathcal{A}^{(-1)}\mathscr{b}^{(i)}$ and a point of accumulation $\mathscr{x}' \in \mathcal{D}$ to this sequence. By the continuity of $\mathcal{A}$ we obtain $\mathcal{A}\mathscr{x}' = \lim_{i\to\infty} \mathscr{b}^{(i)} = \mathscr{b}$. Hence, any convergent subsequence $\{\mathscr{x}^{(i_j)}\}_{j=1}^{\infty}$ fulfils the condition $\lim_{j\to\infty} \operatorname{dist}(\mathscr{x}^{(i_j)}, \mathcal{A}^{(-1)}\mathscr{b}) = 0$. Since $\mathcal{D}$ is compact, this causes $\lim_{i\to\infty} \operatorname{qdist}(\mathcal{A}^{(-1)}\mathscr{b}^{(i)}, \mathcal{A}^{(-1)}\mathscr{b}) = 0$. #

Thus, from the Tikhonov theorem we directly derive stability for all problems the domain $\mathcal{D}$ of which is a bounded subset of a finite dimensional subspace of $B_1$ (see also Remark 2.19). In order to check whether the example problem (2.16) subject to nonnegativity constraints is ill-posed, we now ask for the stability in case of noncompact sets $\mathcal{D}$.

Lemma 2.27:

Under the assumptions of Lemma 2.25 we restrict the problem (2.1) to a noncompact closed convex domain $\mathcal{D}$. Denote by $\bar{H}(\mathcal{D})$ the closure in $B_1$ of the enveloping linear manifold of $\mathcal{D}$, i.e., the closure in

$B_1$ of the smallest linear manifold including $\mathcal{D}$. For given $x^{(0)} \in \mathcal{D}$, we can use the representation $\bar{H}(\mathcal{D}) = \{x \in B_1 : x = x^{(0)} + x', x' \in \tilde{B}_1\}$ with a linear closed subspace $\tilde{B}_1$ of $B_1$, which we assume to be infinite dimensional. In addition, suppose that the relative interior of $\mathcal{D}$ with respect to $\bar{H}(\mathcal{D})$ is nonempty: $\operatorname{int}_{\bar{H}(\mathcal{D})}(\mathcal{D}) \neq \emptyset$. Then the stability condition iii. is injured. o

Proof:
Let $x^{(0)} \in \operatorname{int}_{\bar{H}(\mathcal{D})}(\mathcal{D})$, i.e., there is a sphere $\bar{S}(x^{(0)}, \varepsilon) = \{x \in B_1 : \|x - x^{(0)}\|_1 \leq \varepsilon\}$ with centre $x^{(0)}$ and radius $\varepsilon > 0$ such that we have $\bar{S}(x^{(0)}, \varepsilon) \cap \bar{H}(\mathcal{D}) \subseteq \mathcal{D}$. Lemma 2.25 also applies to the bounded compact injective restriction operator $\tilde{\mathcal{A}} : \tilde{B}_1 \to B_2$ of $\mathcal{A}$. Therefore, $\tilde{\mathcal{A}}_L^{-1} : \mathcal{A}\tilde{B}_1 \to \tilde{B}_1$ is a linear unbounded operator and we obtain infinite sequences $\{x^{(i)}\}_{i=1}^{\infty} \subseteq \tilde{B}_1$ and $\{\mathcal{A}x^{(i)}\}_{i=1}^{\infty} \subseteq B_2$ with $\|x^{(i)}\|_1 = \varepsilon$, $i=1,2,\ldots$ and $\lim_{i\to\infty} \|\mathcal{A}x^{(i)}\|_2 = 0$. This implies $x^{(0)} + x^{(i)} \in \mathcal{D}$, $\|(x^{(0)} + x^{(i)}) - x^{(0)}\|_1 = \varepsilon$, $i=1,2,\ldots$ and $\lim_{i\to\infty} \|\mathcal{A}(x^{(0)} + x^{(i)}) - \mathcal{A}x^{(0)}\|_2 = 0$. Al least at the point $b^{(0)} := \mathcal{A}x^{(0)}$ the continuity requirement iii. of Definition 2.5 fails to be satisfied. #

Remark 2.28:
If the integral operator $\mathcal{A}$ according to (2.16) has a nondegenerate kernel so that $\mathbb{N}(\mathcal{A})$ is trivial, then $\mathcal{A}$ is injective. Since the nonnegativity constraints form a noncompact subset $\mathcal{D}$ of $C[\underline{t}, \bar{t}]$ with $\operatorname{int}(\mathcal{D}) \neq \emptyset$, we can apply Lemma 2.27 in order to show the instability of the considered identification problem. Moreover, $\mathcal{A}$ is never surjective as the following open mapping theorem shows. o

Theorem 2.29:
Let $\mathcal{A}$ be a bounded linear surjective operator, from a Banach space $B_1$ onto a Banach space $B_2$. Then $\mathcal{A}\mathcal{U}$ is an open subset of $B_2$ whenever $\mathcal{U}$ is an open subset of $B_1$. o
Proof: BROWN/PAGE [39, p.317f].

Lemma 2.30:
Under the assumptions of Lemma 2.25 we obtain $\mathbb{R}(\mathcal{A}) \neq B_2$. o

Proof:
The open mapping theorem points out that the inverse mapping $\mathcal{A}_L^{-1}$ of $\mathcal{A}$ is continuous if $\mathcal{A}$ is surjective. Namely, any open set $\mathcal{U} \subseteq B_1$ corresponds to an open set $\mathcal{A}\mathcal{U} \subseteq B_2$ if and only if $\mathcal{A}_L^{-1}$ is a continuous mapping (see DIEUDONNÉ [91, p.55]). However, this is a contradiction as Lemma 2.25 has proven. #

**Corollary 2.31:**
The identification problem (2.16) subject to nonnegativity constraints is an ill-posed one. ○

We are now going to study the discretized version (2.17) through (2.19) of problem (2.16). Assume that A has full column-rank as follows:

$$m \geqslant \operatorname{rank}(A) = n \geqslant 3 \; . \tag{2.22}$$

Then we can find lower bounds for the condition number $\operatorname{cond}_1(A) := \|A\| \cdot \|A^+\|$ that show, for a sufficiently large dimension number n, the ill-conditioning of the problem.

**Theorem 2.32:**
Let the kernel k(s,t) of Eq. (2.16) and its partial derivative $\frac{\partial k(s,t)}{\partial t}$ be continuous functions over the rectangle $P := [\underline{s}, \bar{s}] \times [\underline{t}, \bar{t}]$ with $k'_{max} := \max_P \left|\frac{\partial k(s,t)}{\partial t}\right| > 0$. Provided (2.22) we have

$$\operatorname{cond}_1(A) \geqslant \frac{\sqrt{2} \cdot \underline{k} \cdot n}{k'_{max} \cdot (\bar{t} - \underline{t})} \; . \tag{2.23}$$

If the second partial derivative $\frac{\partial^2 k(s,t)}{\partial t^2}$ is also continuous over P with $k''_{max} := \max_P \left|\frac{\partial^2 k(s,t)}{\partial t^2}\right| > 0$, then

$$\operatorname{cond}_1(A) \geqslant \frac{\sqrt{6} \cdot \underline{k} \cdot n^2}{2 \cdot (\bar{t} - \underline{t})^2 \cdot k''_{max}} \cdot \; ○ \tag{2.24}$$

**Proof:**
Immediately, we obtain $\sup_{0 \neq x \in R^n} \frac{\|Ax\|}{\|x\|} \geqslant \|Ax^{(1)}\| \geqslant \sqrt{m} \cdot \underline{k} \cdot h$ if $x^{(1)} := (0, ., 1, ., 0)^T$

Moreover, for $x^{(2)} := \left(\frac{1}{\sqrt{2}}, \frac{-1}{\sqrt{2}}, 0, 0, \ldots, 0\right)^T$, $\inf_{0 \neq x \in R^n} \frac{\|Ax\|}{\|x\|} \leqslant \|Ax^{(2)}\| =$

$$\left\| \frac{h}{\sqrt{2}} \begin{pmatrix} k(s_1,t_1) - k(s_1,t_2) \\ \vdots \quad\quad \vdots \\ k(s_m,t_1) - k(s_m,t_2) \end{pmatrix} \right\| = \frac{h^2}{\sqrt{2}} \cdot \left\| \begin{pmatrix} \frac{\partial k}{\partial t}(s_1, \eta_1) \\ \vdots \\ \frac{\partial k}{\partial t}(s_m, \eta_m) \end{pmatrix} \right\| \leqslant \frac{h^2}{\sqrt{2}} \sqrt{m} \cdot k'_{max} \text{ with } t_1 < \eta_j < t_2,$$

$j = 1, 2, \ldots, m$.

Thus, $\operatorname{cond}_1(A) \geqslant \frac{\sqrt{m} \cdot \underline{k} \cdot h}{\frac{h^2}{\sqrt{2}} \sqrt{m} \cdot k'_{max}} = \left(\frac{\sqrt{2} \cdot \underline{k}}{k'_{max} \; (\bar{t} - \underline{t})}\right) \cdot n$ . Furthermore, if k is twice differentiable with respect to t, then $x^{(3)} := \left(\frac{1}{\sqrt{6}}, \frac{-2}{\sqrt{6}}, \frac{1}{\sqrt{6}}, 0, , 0\right)^T$ provides $\inf_{0 \neq x \in R^n} \frac{\|Ax\|}{\|x\|} \leqslant \|Ax^{(3)}\| \leqslant \frac{2}{\sqrt{6}} h^3 \cdot k''_{max} \cdot \sqrt{m}$ . This yields the inequality (2.24). #

Analogous theorems could be proven if higher partial derivatives $\frac{\partial^\nu}{\partial t^\nu} k(s,t)$ are continuous functions, where $\text{cond}_1(A) \geq n^\nu \cdot \text{const}$. Consequently, for $k(s,t) \in C^{(\infty)}(P)$, the condition number tends to infinity faster than any power of n. The smoother the kernel, the faster the condition number grows to infinity with n. The same phenomenon (see the coming section) arises in Hilbert space analysis of ill-posed problems.

### 2.2.2. Linear Volterra Integral Equations of the First Kind in Hilbert Spaces

Another class of linear inverse problems (2.1) is formed by Volterra integral equations of the first kind

$$\int_0^s k(s,t)\cdot x(t)\cdot dt = b(s), \quad 0 \leq s \leq 1 \tag{2.25}$$

with a kernel $k(s,t)$ continuous over the triangle $0 \leq t \leq s \leq 1$ (see e.g. BAKER [20], LAVRENTEV [265] and LINZ [277]). We consider the equation (2.25) as an unconstrained identification problem in the Hilbert space $B_1 := B_2 := L_2[0,1]$. Frequently, the interpretation of indirect measurements leads to the special case

$$\int_0^s x(t)\cdot dt = b(s), \quad 0 \leq s \leq 1. \tag{2.26}$$

The solution of Eq. (2.26) requires to find the derivative $x(t)$ of an observed function $b(s)$, where $b(0)=0$. Volterra integral equations occur e.g. in mechanics (find the acceleration of a point mass by observing its velocity), in electrotechnics (see GLASKO [151]) or in rheology (see FRIEDRICH/HOFMANN [133]). Before considering the concrete Eqs. (2.25) and (2.26), we still summarize some main properties of compact operators in Hilbert spaces. For the proofs of the following lemmas, see SCHOCK [386].

In the sequel, let $H_1$, $H_2$ and $H_3$ be real separable Hilbert spaces with $(.,.)_1$, $(.,.)_2$ and $(.,.)_3$ the associated inner products that generate the norms $\|.\|_1$, $\|.\|_2$ and $\|.\|_3$, respectively.

Definition 2.33:
A linear operator $\mathcal{A} : H_1 \to H_2$ is called Hilbert-Schmidt operator if, for an arbitrarily chosen complete orthonormal system $\{u^{(i)}\}_{i=1}^{\infty} \subseteq H_1$, $\sum_{i=1}^{\infty} \|\mathcal{A}u^{(i)}\|_2^2 < \infty$. ○

**Lemma 2.34:**

Any Hilbert-Schmidt operator $\mathcal{A}: H_1 \to H_2$ is compact. $\circ$

**Lemma 2.35:**

The integral operator $\mathcal{A}$ defined by $(\mathcal{A}x)(s) = \int_0^1 k(s,t)x(t)dt$ with a quadratically integrable kernel $\int_0^1 \int_0^1 |k(s,t)|^2 \cdot ds \cdot dt < \infty$ is a Hilbert-Schmidt operator. $\circ$

Lemma 2.35 applies to Volterra integral operators $\mathcal{A}$ according to Eq. (2.25).

**Lemma 2.36:**

Let $\mathcal{A}: H_1 \to H_2$ be a linear compact operator with an infinite dimensional range, i.e., $\dim(\mathcal{R}(\mathcal{A})) = \infty$. Then, there are a sequence of positive real numbers (singular values) $\sigma_1 \geq \sigma_2 \geq \ldots \geq \sigma_i \geq \sigma_{i+1} \geq \ldots$ tending to zero as $i \to \infty$, and a pair of associated orthonormal systems $\{u^{(i)}\}_{i=1}^{\infty} \subseteq H_1$, $\{v^{(i)}\}_{i=1}^{\infty} \subseteq H_2$ such that

$$\mathcal{A}x = \sum_{i=1}^{\infty} \sigma_i \cdot (x, u^{(i)})_1 \cdot v^{(i)}, \tag{2.27}$$

where, for all i,

$$\mathcal{A}u^{(i)} = \sigma_i \cdot v^{(i)}, \quad \mathcal{A}^* v^{(i)} = \sigma_i \cdot u^{(i)}, \|\mathcal{A}\| := \sup_{0 \neq x \in H_1} \frac{\|\mathcal{A}x\|_2}{\|x\|_1} = \sigma_1. \tag{2.28}$$

Here, $\mathcal{A}^*$ denotes the Hilbert adjoint operator of $\mathcal{A}$ defined by the identity $(\mathcal{A}x, b)_2 = (x, \mathcal{A}^* b)_1$, for all $x \in H_1$, $b \in H_2$. $\circ$

The squares $\sigma_i^2$ of singular values $\sigma_i > 0$ of $\mathcal{A}$ are the eigenvalues of the compact selfadjoint positive linear operators $\mathcal{A}^*\mathcal{A}: H_1 \to H_1$ and $\mathcal{A}\mathcal{A}^*: H_2 \to H_2$, for which the singular functions $\{u^{(i)}\}$ and $\{v^{(i)}\}$, respectively, are eigenelements. Note that $\mathcal{A}$ is said to possess the singular system $\{\sigma_i, \; u^{(i)}, \; v^{(i)}\}$, i=1,2,... . Moreover, the sequence of singular values has zero as the only point of accumulation and $(x, u^{(i)})_1 = 0$, i=1,2,..., implies $x \in \mathcal{N}(\mathcal{A})$.

Specifically, for Volterra integral operators $\mathcal{A}$, one can establish the following characteristic property.

**Theorem 2.37:**

For a Volterra integral operator $\mathcal{A}: L_2[0,1] \to L_2[0,1]$ according to Eq. (2.25), the spectrum consists of the number zero. No other eigenvalues exist. The unconstrained fashion of problem (2.25) is an ill-posed identification problem. $\circ$

**Proof:** [236, p.435].

The uniqueness condition ii. is satisfied, for (2.25) in the unconstrained case, iff $\mathbb{N}(\mathcal{A}) = \{0\}$. Then, the operator $\mathcal{A}^{-1}$ is unbounded, since zero is a spectrum point of $\mathcal{A}$. Hence, the stability condition iii. is injured. This instability also results from Lemma 2.25 directly. Finally, Lemma 2.30 indicates that $\mathbb{R}(\mathcal{A}) \neq L_2[0,1]$. We have $L_2[0,1] = \overline{\mathbb{R}(\mathcal{A})} \oplus \mathbb{N}(\mathcal{A}^*)$. Therefore, $\mathbb{R}(\mathcal{A})$ is dense in $L_2[0,1]$ iff the adjoint operator $\mathcal{A}^*$ is injective. The range of compact linear operators in Hilbert spaces is generally characterized by Picard's lemma:

Lemma 2.38:

Consider the linear unconstrained identification problem according to Eq. (2.1), where $\mathcal{A}: H_1 \rightarrow H_2$ is assumed to be injective and compact. Then, for a given right-hand side $\ell$, there is a solution of (2.1) if and only if the Picard condition

$$\sum_{i=1}^{\infty} \frac{(\ell, v^{(i)})_2^2}{\sigma_i^2} < \infty \tag{2.29}$$

is satisfied. ○

Proof:

From the expansion (2.27) it follows $\sum_{i=1}^{\infty} \sigma_i \cdot (\varkappa, u^{(i)})_1 \cdot v^{(i)} = \sum_{i=1}^{\infty} (\ell, v^{(i)})_2 \cdot v^{(i)}$. The orthogonality of $\{v^{(i)}\}_{i=1}^{\infty}$ implies $(\varkappa, u^{(i)})_1 = \frac{(\ell, v^{(i)})_2}{\sigma_i}$, for all i. Hence, there is a solution element $\varkappa \in H_1$ to problem (2.1) iff $\sum_{i=1}^{\infty} (\varkappa, u^{(i)})_1^2 = \sum_{i=1}^{\infty} \frac{(\ell, v^{(i)})_2^2}{\sigma_i^2}$ takes on a finite value. #

Note that $\mathbb{R}(\mathcal{A})$ is the smaller, the faster $\sigma_i$ tends to zero as $i \rightarrow \infty$. Now we are going to specify the general propositions on compact and Volterra integral operators to the linear problem (2.26).

Remark 2.39:

It is not difficult to derive that $\{\sigma_i, u^{(i)}, v^{(i)}\}$ i=1,2,... with

$$\sigma_i = \frac{2}{(2i-1)\pi}, \quad u^{(i)} = \langle u^{(i)}(t) = \sqrt{2} \cdot \cos(i-\tfrac{1}{2})\pi t, \quad 0 \le t \le 1 \rangle, \quad v^{(i)} = \langle v^{(i)}(s) = \sqrt{2} \cdot \sin(i-\tfrac{1}{2})\pi s, \quad 0 \le s \le 1 \rangle \tag{2.30}$$

forms a complete singular system of the Volterra integral operator $\mathcal{A}$ according to (2.26). The considered operator $\mathcal{A}$ transforms a function into its primitive. We see that $\|\mathcal{A}\| = \sigma_1 = \frac{2}{\pi} < 1$, $\sigma_i = O(\frac{1}{i})$ and $\sum_{i=1}^{\infty} \|\mathcal{A} u^{(i)}\|_2^2 = \sum_{i=1}^{\infty} \sigma_i^2 < \frac{1}{\pi^2}(4 + \sum_{i=1}^{\infty} \frac{1}{i^2}) < \infty$. Therefore, $\mathcal{A}$ is a

Hilbert-Schmidt operator of $L_2[0,1]$ into itself. The null-spaces $\mathbb{N}(\mathcal{A})$ and $\mathbb{N}(\mathcal{A}^*)$ with $(\mathcal{A}^*\mathscr{b})(t) = \int_t^1 b(s)ds$, $\mathcal{A}^* : L_2[0,1] \to L_2[0,1]$ are trivial and the range $\mathbb{R}(\mathcal{A})$, dense in $L_2[0,1]$, is formed by the family of absolutely continuous functions $b(s)$, $0 \le s \le 1$, $b(0)=0$, the generalized derivative of which is of $L_2[0,1]$ (see [236, p.64]). Thus, $\mathbb{R}(\mathcal{A})$ is a subset of the Sobolev space $W_2^{(1)}[0,1]$ of functions $\mathscr{b} = \langle b(s),\ 0 \le s \le 1,\ b'(s) \in L_2[0,1] \rangle$ having a quadratically integrable generalized derivative $b'(s)$ (see SMIRNOV [402, Vol. V]). Here,

$$\|\mathscr{b}\|_{W_2^{(1)}[0,1]} := \left( \int_0^1 b^2(s)ds + \int_0^1 (b')^2(s)ds \right)^{1/2} \tag{2.31}$$

forms the norm in the Hilbert space $W_2^{(1)}[0,1]$. o

Lemma 2.40:

If $\mathcal{A}$ corresponds to Eq. (2.26), then there are real constants $0 < \underline{c} \le \overline{c}$ so that, for all $\varkappa \in L_2[0,1]$,

$$\underline{c} \cdot \|\varkappa\|_{L_2[0,1]} \le \|\mathcal{A}\varkappa\|_{W_2^{(1)}[0,1]} \le \overline{c} \cdot \|\varkappa\|_{L_2[0,1]} . \text{ o} \tag{2.32}$$

Proof:

We have $\|\mathcal{A}\varkappa\|^2_{W_2^{(1)}[0,1]} = \|\mathcal{A}\varkappa\|^2_{L_2[0,1]} + \|\varkappa\|^2_{L_2[0,1]}$. From Remark 2.39 it follows $\|\mathcal{A}\varkappa\|_{L_2[0,1]} \le \frac{2}{\pi} \cdot \|\varkappa\|_{L_2[0,1]}$. Thus, the inequality (2.32) is valid whenever we set $\underline{c} := 1$ and $\overline{c} := \sqrt{1 + \frac{4}{\pi^2}}$ . #

From the Theorems, Lemmas and Remarks 2.25, 2.29, 2.39 and 2.40 we obtain:

Corollary 2.41:

Let $\mathcal{A} : L_2[0,1] \to L_2[0,1]$ be the integral operator of Eq. (2.26). Then, $\mathbb{R}(\mathcal{A})$ is a subspace of $W_2^{(1)}[0,1]$. The mapping $\mathcal{A} : L_2[0,1] \to \mathbb{R}(\mathcal{A}) \subseteq W_2^{(1)}[0,1]$ and its inverse mapping $\mathcal{A}_L^{-1}$ are continuous. If we define the inner product on $\mathbb{R}(\mathcal{A})$ by $(\mathcal{A}\varkappa, \mathcal{A}y)_{\mathbb{R}(\mathcal{A})} := (\varkappa, y)_{L_2[0,1]}$, then $H := \mathbb{R}(\mathcal{A})$ gets a Hilbert space, where $\mathcal{A} : L_2[0,1] \to H$ is even bijective. Thus, the problem (2.26) satisfies the uniqueness and stability conditions ii. and iii. if $H_1 := L_2[0,1]$, $H_2 := W_2^{(1)}[0,1]$. However, it is even well-posed if $H_1 := L_2[0,1]$, $H_2 := H$. Note that, for $\mathcal{A} : L_2[0,1] \to W_2^{(1)}[0,1]$, $(\mathcal{A}^*\mathscr{b})(t) = \int_t^1 b(s)ds + b'(t)$, $\dim(\mathbb{N}(\mathcal{A})) = 1$ and the function $\mathscr{b} = \langle b(s) = \cosh(1-s),\ 0 \le s \le 1 \rangle \in \mathbb{N}(\mathcal{A}^*)$. o

Considering our linear equation (2.26), an instable ill-posed inverse problem is transformed into a stable or well-posed one by using a stronger norm for evaluating the space of observable elements (see Corollary 2.41). The same effect could be achieved if the right-hand side space norm would remain as before, but a weaker norm should operate on the unknown elements. However, this way to get stability and well-posedness is not practicable for applied problems, since the norms and admissible sets for the right-hand side are determined by the physical character of measurements and unknown objects.

The level of difficulties in solving linear unconstrained identification problems with compact operators $\mathcal{A}$ in Hilbert spaces may be expressed by the degree of ill-posedness. This degree is given by the rate of convergence of the singular values $\sigma_i$ to zero as $i \to \infty$. The faster $\sigma_i$ becomes small, the more the problem is instable. Another strategy of determining the degree of ill-posedness uses norm equivalences similar to formula (2.32). We discuss the interrelations between these two ways of finding an instability measure. A comparison between the linear problem (2.26) and ill-posed neighbouring problems with respect to the degree of ill-posedness will close the investigations of this section.

Definition 2.42:
Let (2.1) be a linear unconstrained identification problem, where $\mathcal{A}$ satisfies the assumptions of Lemma 2.36. Then, a positive real number $\vartheta$ is called the degree of ill-posedness to this problem if, for the singular values $\sigma_i$ of $\mathcal{A}$, we have $\sigma_i = O(\frac{1}{i^\vartheta})$. We say that (2.1) is mildly ill-posed if $\vartheta \leq 1$, moderately ill-posed if $1 < \vartheta < \infty$. On the other hand, the case $\sigma_i = o\,(\frac{1}{i^\vartheta})$, for all $\vartheta > 0$, i.e., $\vartheta = \infty$, is termed severly ill-posed. $\circ$

WAHBA [470] has presented the concept of mildly, moderately and severely ill-posed problems in another way based on the "intrinsic rank" of a semi-discretized problem. WAHBA's classification of ill-posed problems and the classification of Definition 2.42 are very similar.

Now consider a Galerkin method discretization to problem (2.1) in Hilbert spaces with compact operator $\mathcal{A}$. We suppose

$$\begin{aligned} P^n \varkappa &:= x = ((\varkappa, e^{(1)})_1, (\varkappa, e^{(2)})_1, \ldots, (\varkappa, e^{(n)})_1)^T \\ Q^m \ell &:= b = ((\ell, f^{(1)})_2, (\ell, f^{(2)})_2, \ldots, (\ell, f^{(m)})_2)^T \qquad (2.33) \\ \wp^n x &:= \sum_{i=1}^{n} x_i \cdot e^{(i)} \end{aligned}$$

and a system matrix

$$A = \left( (\mathcal{A}e^{(i)}, f^{(j)})_2 \right)_{\substack{j=1(1)m \\ i=1(1)n}}, \quad \text{rank}(A)=n, \tag{2.34}$$

defining a discretized identification problem (2.4). Again, $\{e^{(i)}\}_{i=1}^{\infty}$ and $\{f^{(j)}\}_{j=1}^{\infty}$ denote complete orthonormal systems of $H_1$ and $H_2$. Owing to Schwarz's inequality, the discretization error $\|Q^m\mathcal{A}\varkappa - AP^n\varkappa\|^2 = \sum_{j=1}^{m} \left( \sum_{i=n+1}^{\infty} (\varkappa, e^{(i)})_1 (\mathcal{A}e^{(i)}, f^{(j)})_2 \right)^2$ tends to zero as $n \to \infty$.

Theorem 2.43:

The linear discretized identification problem (2.4) according to (2.33) and (2.34) satisfies the inequality

$$\text{cond}_1(A) \geqslant \frac{a_{max}^{m,n}}{\min\limits_{0 \neq \varkappa \in \text{span}(e^{(1)},\dots,e^{(n)})} \frac{\|\mathcal{A}\varkappa\|_2}{\|\varkappa\|_1}} \geqslant \frac{a_{max}^{m,n}}{\sigma_n}, \tag{2.35}$$

where $a_{max}^{m,n} := \max\limits_{\substack{1 \leq j \leq m \\ 1 \leq i \leq n}} |(\mathcal{A}e^{(i)}, f^{(j)})_2|$ and $\text{span}(e^{(1)},\dots,e^{(n)})$ designates the subspace of $H_1$ generated by the first n basis functions of $\{e^{(i)}\}$. ○

In order to prove Theorem 2.43, we mention a corollary to the Poincaré-Fischer extremum principle (see e.g. [386]).

Lemma 2.44:

Under the assumptions of Lemma 2.36 we have, for all i=1,2,...,

$$\sigma_i = \max_{H^{(i)} \subseteq H_1} \min_{0 \neq \varkappa \in H^{(i)}} \frac{\|\mathcal{A}\varkappa\|_2}{\|\varkappa\|_1} = \min_{0 \neq \varkappa \in \text{span}(u^{(1)},\dots,u^{(i)})} \frac{\|\mathcal{A}\varkappa\|_2}{\|\varkappa\|_1}, \tag{2.36}$$

where $H^{(i)}$ represents an arbitrary subspace of $H_1$ with $\dim(H^{(i)}) = i$. ○

Proof of Theorem 2.43:

We can see that, for a certain vector $x^{(0)} := (0,..,0,1,0,..,0)^T$, $\sup\limits_{0 \neq x \in R^n} \frac{\|Ax\|}{\|x\|} \geq \|Ax^{(0)}\| \geq a_{max}^{m,n}$. Furthermore, we obtain $(Ax)_j = (\mathcal{A}P^n x, f^{(j)})_2$ and

$$\min_{0 \neq x \in R^n} \frac{\|Ax\|^2}{\|x\|^2} = \min_{0 \neq x \in R^n} \frac{\sum_{j=1}^{m} (\mathcal{A}P^n x, f^{(j)})_2^2}{\|P^n x\|_1^2} \leq \min_{0 \neq x \in R^n} \frac{\|\mathcal{A}P^n x\|_2^2}{\|P^n x\|_1^2} =$$

$$= \min_{0 \neq \varkappa \in \text{span}(e^{(1)},\dots,e^{(n)})} \frac{\|\mathcal{A}\varkappa\|_2^2}{\|\varkappa\|_1^2}.$$

Hence, by definition of the numerical condition number, the left inequality of (2.35) holds. The upper bound obtained for the denominator of $\mathrm{cond}_1(A)$ attains its maximum

$$\sigma_n = \max_{\mathrm{span}(e^{(1)},\ldots,e^{(n)})} \min_{0\neq \varkappa \in \mathrm{span}(e^{(1)},\ldots,e^{(n)})} \frac{\|\mathcal{A}\varkappa\|_2}{\|\varkappa\|_1} \quad \text{if } e^{(i)} = u^{(i)}, \quad i=1(1)n.$$

This direct application of Lemma 2.44 yields the right inequality of (2.35). #

The above given proof suggests choosing the singular functions $\{u^{(i)}\}$, $i=1(1)n$, associated with the largest singular values as a basis for discretising $\varkappa$ in order to keep the numerical condition number small. On the other hand, the basis $\{f^{(j)}\}_{j=1}^{m}$ will frequently be determined by the observation characteristics. If $H_2=L_2[0,1]$, then $(\ell, f^{(j)})_{L_2[0,1]}$ is usually a weighted mean of b(s) over a subinterval $[s_j, s_{j+1}] \subseteq [0,1]$ with mid-point $s_j'$:

$$(\ell, f^{(j)})_2 = \int_{s_j}^{s_{j+1}} f(|s-s_j'|)\cdot b(s)\, ds\ , \tag{2.37}$$

where f(u) is a monotonically nonincreasing function with respect to $u \geq 0$.

Corollary 2.45:

Under the assumptions of Theorem 2.43 consider a sequence of discretised versions (2.33), (2.34) to Eq. (2.1). Then, the associated numerical condition numbers $\mathrm{cond}_1(A)$ tend to infinity as $n \to \infty$ with a minimum rate $\mathrm{cond}_1(A) \propto n^{\vartheta}$ whenever $\vartheta$ is the degree of ill-posedness to problem (2.1). If, however, (2.1) is severely ill-posed, then the condition number grows faster than any power of n. ○

For a Hilbert space problem (2.1), the singular values $\sigma_i$ are rarely available explicitly. Then, we may use a second route to determine the degree of ill-posedness. The following lemma is a direct consequence of Lemma 2.44.

Lemma 2.46:

Let $\mathcal{A}: H_1 \to H_2$ satisfy the assumptions of Lemma 2.36, where $H_1$ is a subset of the Hilbert space $H_3$. Furthermore, let the embedding operator $\mathcal{E}: H_1 \to H_3$ be compact with singular values $\sigma_1' \geq \sigma_2' \geq \ldots \geq \sigma_i' \geq \ldots > 0$. If an inequality of the form

$$\underline{c}\cdot\|\varkappa\|_3 \leq \|\mathcal{A}\varkappa\|_2 \leq \bar{c}\cdot\|\varkappa\|_3\ , \quad 0<\underline{c}\leq\bar{c} \tag{2.38}$$

holds, for all $\varkappa \in H_1$, then we have upper and lower bounds

$$\underline{c}\cdot\sigma_i' \leq \sigma_i \leq \bar{c}\cdot\sigma_i'\ , \quad i=1,2,\ldots\ , \tag{2.39}$$

for the singular values $\sigma_i$ of $\mathcal{A}: H_1 \to H_2$. ○

Referring back to NATTERER [340] (see also ENGL/NEUBAUER [117]) we can apply Lemma 2.46 to Hilbert scales. This theory shall be explained by means of a special case, which is important for neighbouring problems of (2.26).

Let $\mathcal{B}$ be the densely defined linear unbounded selfadjoint strictly positive operator on $\tilde{H}^{(0)} := L_2[0,1]$ determined by the normalized square root $\mathcal{B} = \frac{1}{\pi}\tilde{\mathcal{B}}^{1/2}$ of the second order differential operator $(\tilde{\mathcal{B}}x)(t) = -x''(t)$ if zero Dirichlet data are imposed on the boundary. Then, $\lambda_i = i \geq 1$, $u^{(i)}(t) = \sqrt{2}\cdot\sin i\pi t$, $i=1,2,\ldots$, form the eigenvalues and eigenfunctions of $\mathcal{B}$. We introduce a continuous scale of Hilbert spaces $\tilde{H}^{(\mu)}$, $\mu \in R$, where $\tilde{H}^{(\mu)}$ is the completion of $\bigcap_{i=1}^{\infty} \mathcal{D}(\mathcal{B}^i)$ (here, $\mathcal{D}(\mathcal{B}^i)$ denotes the maximum domain of the operator power $\mathcal{B}^i$) with respect to the Hilbert space norm $\|x\|_{\tilde{H}^{(\mu)}} := \|\mathcal{B}^{\mu} x\|_{\tilde{H}^{(0)}}$. Now it is provable that

$\tilde{H}^{(\mu)} = \{x = \sum_{i=1}^{\infty} x_i \cdot u^{(i)} : \sum_{i=1}^{\infty} \lambda_i^{2\mu} \cdot x_i^{2\mu} < \infty\}$ and thus the embedding operator $\mathcal{E}: \tilde{H}^{(\mu)} \to \tilde{H}^{(\nu)}$, $-\infty < \nu < \mu < \infty$ is compact and possesses the singular values $\sigma_i = \lambda_i^{\nu-\mu}$, $i=1,2,\ldots$ . In the considered case $\tilde{H}^{(0)} = L_2[0,1]$, we obtain the Hilbertian Sobolev spaces $\tilde{H}^{(\mu)} = H_0^{(\mu)}[0,1]$ (see NATTERER [335]) defined as follows: $H_0^{(\mu)}[0,1] = \{x(t) \in H^{(\mu)}(R) : \operatorname{supp}(x) \subseteq [0,1]\}$, where $H^{(\mu)}(R)$ designates the set of all tempered distributions satisfying

$$\|x\|_{H^{(\mu)}(R)} := \left(\int_{-\infty}^{\infty} (1+|\omega|^2)^{\mu} \cdot |\hat{x}(\omega)|^2 \cdot d\omega\right)^{\frac{1}{2}} < \infty \tag{2.40}$$

with $\|x\|_{H_0^{(\mu)}[0,1]} = \|x\|_{H^{(\mu)}(R)}$ if $x \in H_0^{(\mu)}[0,1]$. Here, $\hat{x}(\omega)$ is the Fourier transform of $x(t)$: $\hat{x}(\omega) := \frac{1}{\sqrt{2\pi}} \cdot \int_{-\infty}^{\infty} \exp(-i\omega t) \cdot x(t) \cdot dt$. Note that $H_0^{(\mu)}[0,1]$ and $\mathring{W}_2^{(\mu)}[0,1]$ (see [402]) coincide if $\mu$ is a positive integer and we have $H_0^{(0)}[0,1] = L_2[0,1]$. For properties of tempered distributions and of the Fourier transform we refer to LIONS/MAGENES [282, Chap.1 ]. By using tempered distributions one can derive inequalities of the form (2.38).

Corollary 2.47:

Suppose that $\mathcal{A}: L_2[0,1] \to L_2[0,1]$ is linear and the inequality

$$\underline{c}\|x\|_{H_0^{(-\mu)}[0,1]} \leq \|\mathcal{A}x\|_{L_2[0,1]} \leq \bar{c}\|x\|_{H_0^{(-\mu)}[0,1]} \tag{2.41}$$

holds for all $x \in L_2[0,1]$ and fixed values $\mu > 0$, $0 < \underline{c} \leq \bar{c}$. Then, the associated unconstrained identification problem (2.1) possesses a compact operator $\mathcal{A}$ and a degree $\mu$ of ill-posedness. $\circ$

We are now going to use this corollary in order to compare a family of identification problems (2.1) with $\mathfrak{D} := B_1 := B_2 := L_2[0,1]$ by considering their degree of ill-posedness. Based on this comparison, the behaviour of corresponding discretized problems (2.4) according to (2.33), (2.34) is indicated by Theorem 2.43.

Thus, we deal with the problem of finding the l-th derivative $b^{[l]}(s)$ of a sufficiently smooth function $b(s)$, $0 \leq s \leq 1$, satisfying the condition $b^{[\nu]}(0)=0$, $\nu=0,1,\ldots,l-1$ :

$$\int_0^s \frac{(s-t)^{l-1}}{(l-1)!} \cdot x(t) \cdot dt = b(s), \quad 0 \leq s \leq 1, \; l=1,2,\ldots \tag{2.42}$$

(see e.g. WAHBA [470]). These Volterra equations possessing a convolution kernel, for all integers $l \geq 1$, represent special cases of Eq. (2.25). Furthermore, consider an Abel integral equation being mildly singular

$$\int_0^s \frac{x(t)}{\sqrt{s-t}} \cdot dt = b(s), \quad 0 \leq s \leq 1 \tag{2.43}$$

(see e.g. ANDERSSEN [9], GORENFLO [170], for applications of the Abel integral equation in computed tomography CORMACK [75]). Finally, take into consideration an integral equation of Fredholm type with $C^\infty$-kernel

$$\int_0^1 2 \cdot \left(\sum_{i=1}^{\infty} \exp(-i^2\pi^2) \cdot \sin i\pi s \cdot \sin i\pi t\right) \cdot x(t) \cdot dt = b(s), \quad 0 \leq s \leq 1 \tag{2.44}$$

(for the physical background of this equation, see Sec. 2.2.5). Any integral equation of the form (2.42), (2.43) or (2.44) is of importance in mathematical modelling for applied inverse problems.

Well-known properties of Fourier transform and the arguments of [332, Proof of Lemma 4.1] allow us to prove inequalities (2.41) for integral operators according to Eqs. (2.42) and (2.43) on $L_2[0,1]$. We derive that $\mu = l$ for the l-th derivative problem (2.42) and $\mu = \frac{1}{2}$ for the Abel equation (2.43). On the other hand, it may be shown that in case of Eq. (2.44) $\|\mathcal{A}\varkappa\|_{L_2[0,1]}$ is weaker than any norm $\|\varkappa\|_{H_0^{(-\mu)}[0,1]}$, $\mu > 0$ (see [131]). Like in the C-space analysis of our previous paragraph we can point out, here, a dependence between the smoothness of the kernel $k(s,t)$ and the degree of ill-posedness in the Hilbert space $L_2[0,1]$. The smoother the kernel, the more instable the associated linear inverse problem becomes.

The Abel integral equation is mildly ill-posed ($\vartheta = \frac{1}{2}$), whereas differentiation ($\vartheta = 1$) lies on the boundary between mildly and moderately ill-posed problems. The degree of ill-posedness for finding the l-th derivative ($\vartheta = l$) grows with l and shows moderate ill-posedness. Hence, it is difficult to find a higher derivative with a desired accuracy when the given data are noisy. However, the chances to get a convenient solution to (2.42) are good in comparison with those of problem (2.44). The singular values $\sigma_i = \exp(-i^2\pi^2)$ of this severely ill-posed problem rapidly decrease as i increases. A discretized version of such a $C^\infty$-kernel problem is even extremely ill-conditioned if the dimension number n remains very small.

### 2.2.3. The Linear Identification Problem of Computerized Tomography

The previous sections have presented a study on the linear problem of identifying a function on an interval. Now we are going to leave the one-dimensional case and turn to a linear problem from picture reconstruction. Since computerized tomography (for short CT) was invented by HOUNSFIELD and CORMACK (see [75] for a historical review of CT, [C19] for mathematical aspects), considerable attention has been paid to picture reconstruction from projections (see HERMAN [C14]).

In the general picture reconstruction problem, a picture density $\mathcal{X} = \langle x(t) \geq 0,\ t = (t_1, t_2)^T \in R^2,\ \mathrm{supp}(x) \subseteq G \rangle$ vanishing outside a given bounded domain G of the two dimensional space is to be determined from a family of integrals

$$f_\Lambda(\mathcal{X}) = \int_\Lambda x(t) \cdot dt \ . \tag{2.45}$$

Here, $\Lambda$ is a straight line in $R^2$ crossing the domain G. Specifically, computerized tomography tries to reconstruct, for diagnostic purposes in medicine, the two-dimensional density distribution of a human body along a fixed slice plane through the body. We get data of the form (2.45) by means of a scanner that measures the attenuation of X-rays along a line lying in this slice. In the sequel, assume the density to be concentrated in the unit circle $G := \{ t \in R^2 : \|t\| \leq 1 \}$. We denote by $\Sigma$ the boundary of G.

Speaking in mathematical terms (see NATTERER [335], also [286]), the problem of CT can be expressed by the following linear equation

$$\int_{\Lambda = \{t \in R^2 :\ (t,\omega) = s\}} x(t) \cdot dt = b(s,\omega), \quad s \in R,\ \omega = (\cos\varphi, \sin\varphi)^T \in R^2, \quad 0 \leq \varphi < 2\pi . \tag{2.46}$$

Consider this linear identification problem in the Hilbert spaces $B_1 := L_2(G)$ and $B_2 := L_2(Z)$, where $Z := R \times \Sigma \subseteq R^3$ is the unit cylinder surface generated by $\Sigma$. Moreover, for $\mathcal{b} = \langle b(s,\omega), s \in R, \omega \in \Sigma \rangle$, let

$$\|\mathcal{b}\|_{L_2(Z)} := \left( \int_{-\infty}^{\infty} \int_0^{2\pi} (b(s,\cos\varphi,\sin\varphi))^2 \cdot d\varphi\, ds \right)^{1/2} \tag{2.47}$$

be the norm of $B_2$. In Eq. (2.46), the scalar quantity s describes the perpendicular distance of the observed line $\Lambda$ from the origin. On ther other hand, $\varphi$ expresses the associated angle.

Thus, Eq. (2.46) is a special linear identification problem (2.1), the operator $\mathcal{R}$ of which is the so-called Radon transform. Therefore, the solution of the CT-integral equation is equivalent to the inversion of Radon transform. An extensive survey of numerical methods for solving (2.46) is given in [339], (see also [333]). We will outline a discretized version to (2.46) below. However, beforehand let us summarize the main properties of the Radon transform $\mathcal{R}$.

Lemma 2.48:
For the Radon transform $\mathcal{R}$ according to Eq. (2.46) we have $0 < \underline{c} \leq \bar{c}$ such that

$$\underline{c} \cdot \|\varkappa\|_{H_0^{(-\frac{1}{2})}(G)} \leq \|\mathcal{R}\varkappa\|_{L_2(Z)} \leq \bar{c} \cdot \|\varkappa\|_{H_0^{(-\frac{1}{2})}(G)} \tag{2.48}$$

(as for the definition of $H_0^{(-\frac{1}{2})}(G)$ see [335] ). $\circ$

Proof: [332, Lemma 4.1].

By recalling Lemma 2.46 and Definition 2.42 we obtain:

Corollary 2.49:
The CT-integral equation (2.46) with $B_1 = L_2(G)$, $B_2 = L_2(Z)$ is a mildly ill-posed problem possessing a degree of ill-posedness $\vartheta = \frac{1}{2}$. From (2.48) it follows that the associated linear compact operator $\mathcal{R} : B_1 \to B_2$ is injective over $B_1$. However, $\mathcal{R}$ fails to be surjective since the range $\mathcal{R}(\mathcal{R})$ is a dense but nonclosed subset of $B_2$ with respect to the norm (2.47). The inverse mapping $\mathcal{R}_L^{-1}$ : $B_2 \to B_1$ is unbounded. $\circ$

Note that the problem (2.46) becomes well-posed if the solution space norm is weakened to $B_1 := H_0^{(-\frac{1}{2})}(G)$. However, this does not facilitate the numerical solution of CT-problems. Namely, norms weaker than $L_2(G)$ are not of practical relevance as measures for the precision of a solution. Furthermore, note that the ill-posedness of Eq. (2.46) is maintained if nonnegativity constraints are added (see Sec. 2.2.1.).

Now let us discretize Eq. (2.46) by using the formulae (2.33), (2.34) with basis functions $e^{(1)},\ldots,e^{(n)}$ for the solution and $f^{(1)},\ldots,f^{(m)}$ for the right-hand side. Here, we subdivide the unit square $\tilde{G}$ containing G into $n:=n_0^2$ subsquares $G_i$ with $\tilde{G}=\bigcup_{i=1}^{n} G_i$ and define $e^{(i)}:=\left\langle e^{(i)}(t)=\begin{cases} n_0 & \text{if } t\in G_i \\ 0 & \text{if } t\notin G_i \end{cases}\right\rangle \in L_2(\tilde{G})$.

On the other hand, we pick $m:=m_1\cdot m_2$ observation data for $m_1$ directions $\varphi_1,\ldots,\varphi_{m_1}$ and $m_2$ values $s_1,\ldots,s_{m_2}$ for the perpendicular distance of a line $\Lambda$ from the origin. The corresponding basis functions $f^{(1)},\ldots,f^{(m)}$ are chosen similar to (2.37) with respect to the three-dimensional character of Z. We have to observe integrals

$$(b,f^{(j)})_{L_2(Z)}=\int_{-1}^{1}\int_{0}^{2\pi} f(|s-s_{j_2}|,|\varphi-\varphi_{j_1}|)\cdot b(s,\cos\varphi,\sin\varphi)\cdot d\varphi\, ds \tag{2.49}$$

with $s_{j_2}$ and $\varphi_{j_1}$ distance and direction associated with the j-th observation and with a function f(u,v) which is monotonically nonincreasing with respect to $u\geq 0$ and $v\geq 0$.

The discrete version (2.4) to (2.46) subject to the discretization described above is characterized by two main properties. Firstly, if the measurement tool function f(.,.) of Eq. (2.49) is close to a $\delta$-distribution, then the system matrix A gets sparse. The finer the grid over $\tilde{G}$, the more entries $a_{ji}$ of A will be zero. Really, any straight line $\Lambda$ generating a row of the matrix A can cross not more than $n_0$ subsquares $G_i$ of $\tilde{G}$. Secondly, the two-dimensional geometry requires, for a practically acceptable picture reproduction, a number of data points and subsquares $G_i$ that can be in the many thousands. This forces to solve large linear systems with a system matrix $A\in\mathbb{R}^{m_1\cdot m_2\times n_0^2}$ (see e.g. [470]). The condition number behaviour results from Theorem 2.43 and indicates a minimum rate of $\mathrm{cond}_1(A)\sim\sqrt{n}$ as $n\to\infty$. It is fortunate that the problem is only very mildly ill-posed.

## 2.2.4. A Nonlinear Identification Problem of Convolution Type

This is the first example of a nonlinear identification problem (2.1). We consider (see also [206]) the convolution of a nonnegative function by itself in the spaces $B_1:=B_2:=C[0,1]$:

$$\int_0^s x(s-t)\cdot x(t)\cdot dt = b(s)\ ,\quad 0\leq s\leq 1\ . \tag{2.50}$$

**Theorem 2.50:**

The integral operator $\mathcal{A}: \mathcal{D} \subseteq B_1 \longrightarrow B_2$ according to Eq. (2.50) and $\mathcal{D} := \{ \varkappa \in B_1 : x(t) \geq 0,\ 0 \leq t \leq 1 \}$ is nonlinear and continuous. Any of the requirements i., ii. and iii. of the well-posedness definition is injured. o

**Proof:**

The operator $\mathcal{A}$ is obviously nonlinear. Its continuity results from the estimate

$$\left| \int_0^s x^{(0)}(s-t) \cdot x^{(0)}(t) \cdot dt - \int_0^s x(s-t) \cdot x(t) \cdot dt \right| \leq \int_0^s |x^{(0)}(s-t) - x(s-t)| \cdot$$

$$\cdot |x^{(0)}(t)|\, dt + \int_0^s |x(s-t)| \cdot |x^{(0)}(t) - x(t)| \cdot dt$$ . Hence, $\| \mathcal{A}\varkappa^{(0)} - \mathcal{A}\varkappa \|_2 \leq$

$\| \varkappa^{(0)} - \varkappa \|_1 \cdot ( \| \varkappa^{(0)} \|_1 + \| \varkappa \|_1 )$ and $\| \mathcal{A}\varkappa^{(0)} - \mathcal{A}\varkappa \|_2 \to 0$ as $\| \varkappa^{(0)} - \varkappa \|_1 \to 0$.

Some counterexamples will disprove the well-posedness of (2.50). We easily disprove the existence condition i. if considering functions $\ell = \langle b(s),\ 0 \leq s \leq 1 \rangle$, which are not monotonically nondecreasing. Namely, owing to $x(t) \geq 0$, $0 \leq t \leq 1$, the values $b(s)$ must not fall as $s$ grows. Furthermore, any nonnegative continuous function $\varkappa = \langle x(t),\ 0 \leq t \leq 1 \rangle$ with $\mathrm{supp}(x) \subseteq [\frac{1}{2}, 1]$ causes a zero right hand side $\ell = \mathcal{A}\varkappa = 0 = \langle b(s) \equiv 0,\ 0 \leq s \leq 1 \rangle$. Thus, $\mathcal{A}$ fails to be injective over $\mathcal{D}$. Finally, a sequence

$$\varkappa^{(i)} := \left\langle x^{(i)}(t) = \begin{cases} 1 & \text{if } 0 \leq t < 1 - \frac{1}{i} \\ 2 + i(t-1) & \text{if } 1 - \frac{1}{i} \leq t \leq 1 \end{cases} \right\rangle, \quad i = 2, 3, \ldots$$

belongs to $\mathcal{D}$. However, it is evident that $x^{(i)}(1) = 2$ and one shows easily that, for $\varkappa^{(0)} = \langle x^{(0)}(t) \equiv 1,\ 0 \leq t \leq 1 \rangle$, $\| \varkappa^{(i)} - \varkappa^{(0)} \|_1 = 1$ and $\| \mathcal{A}\varkappa^{(i)} - \mathcal{A}\varkappa^{(0)} \|_2$ tends to zero as $i \to \infty$. Not even the inverse operator $\mathcal{A}^{(-1)}$ is a continuous multi-valued mapping. #

**Remark 2.51:**

We mention that the finiteness of the domain [0,1] for the functions $\varkappa$ and $\ell$ leads to rising difficulties compared to a modified case of Eq. (2.50) in which the domain is extended to the half-axis $0 \leq s < \infty$ (see RICHTER [369]). If we were studying this equation in $B_1 := B_2 := C[0, \infty)$, then Laplacian transform would help to show the injectivity of $\mathcal{A}$ over $\mathcal{D}$. However, we would need more input data and hence more function values $b(s)$ with $s > 1$. o

The present nonlinear convolution problem is of practical importance in mathematical statistics (see [369]). If $X_\tau(\omega)$ represents a Gaussian random process with time $\tau$, zero mean and given correlation function, then the distribution function of the random variable

$X(\omega) := \int_0^T X_\tau^2(\omega)\cdot d\tau$ on the subinterval [0,1] can be identified via Eq. (2.50). Here, the function $\mathscr{b}$ is available by the moments of $X_\tau(\omega)$.

We are now going to discretise the problem. The projections $P^n$ and $Q^m$ are chosen analogously to (2.17) and (2.18). However, the structure of $\mathscr{A}$ suggests equating m with n:

$$t_i := \frac{2i-1}{2n}, \quad s_j := \frac{j}{n}, \quad i=1(1)n, \quad j=1(1)n,$$
$$P^n \varkappa := x = (x(t_1),\ldots,x(t_n))^T \in R^n, \quad Q^m \mathscr{b} := b = (b(s_1),\ldots,b(s_n))^T \in R^n. \tag{2.51}$$

So it is possible to apply the rectangular rule as follows:

$$(Ax)_j := \frac{1}{n}\cdot \sum_{i=1}^{j} x_{j+1-i}\cdot x_i, \quad j=1(1)n. \tag{2.52}$$

The nonlinear operator A: $R^n \to R^n$ can be represented by

$$Ax := \frac{1}{n}\cdot M(x)\cdot x \quad , \text{ where } M(x) := \begin{bmatrix} x_1 & & & 0 \\ x_2 & \ddots & & \\ \vdots & \ddots & \ddots & \\ x_n & \cdots & x_2 & x_1 \end{bmatrix}. \tag{2.53}$$

Thus, range vectors Ax, containing powers of $x_i$ up to the second order only, are formed by a product of a lower triangular Toeplitz matrix M(x) and a multiple of the vector x. Note that the Fréchet derivative $A'(x)$ of A at a point x is the matrix $A'(x) = \frac{2}{n}\cdot M(x)$. Whenever we treat the discretised fashion Ax=b, for this example, we have to accept nonnegativity constraints $P^n\mathscr{D} = D = \{x \in R^n: x_i \geq 0, \ i=1(1)n\}$ resulting from the nonnegativity of the continuous background problem.

Lemma 2.52:

The unconstrained problem (2.4) according to (2.51) and (2.52) is uniquely solvable if the first component $b_1$ of the right-hand side vector does not vanish. ○

Proof:

The triangular structure of A allows the successive verification of $x_1, x_2, \ldots, x_n$. If we already have calculated $x_1, \ldots, x_{\nu-1}$ by means of the first $\nu-1$ equations of Ax=b, then we can determine $x_\nu$ with $\nu \leq n$ from the $\nu$-th equation in a unique manner whenever $x_1 \neq 0$. Thus, we derive

$$x_1 = \sqrt{n\cdot b_1}\ ; \ x_2 = \frac{n\cdot b_2}{2\cdot x_1}\ ; \ \ldots\ ; \ x_j = \frac{n\cdot b_j - \sum_{i=2}^{j-1} x_{j+1-i}\cdot x_i}{2\cdot x_1}, \quad j=3(1)n. \tag{2.54}$$

Due to this explicit verification the lemma proved to be correct. ✱

Corollary 2.53:
The discrete version (2.4) of (2.50) according to the formulae (2.51) and (2.52) is ill-posed subject to nonnegativity constraints. o

The assertion of this corollary is a consequence of the formula (2.54). A unique solution of $Ax=b$ exists if $b_1 \neq 0$, however, the components $x_i$ may be negative even if $b_j > 0$, for all $j=1(1)n$. Provided $b_1=0$, the system has only the trivial solution $x=0$ if $b=0$. Otherwise, for $b_1=0$ and $b_j \neq 0$ with a certain subscript $j$ and $2 \leq j \leq n$, the system is inconsistent. Thus, A is injective but not surjective over D. For very small values $b_1$, the component $x_1$ is also small. Consequently, the solution vector x tends to be strongly oscillating in its components between absolutely large positive and negative values (see the numerical example in [206]). The condition number $cond_1(.)$ gives a deeper insight into the stability behaviour of the problem and into the distinction between linear and nonlinear inverse problems.

We present a further well-known lemma from numerical analysis concerning nonlinear identification problems (see also Sec. 2.1., Definition 2.13). For the proof of this lemma see [364].

Lemma 2.54:
Let (2.4) be a nonlinear identification problem with a continuous operator $A: D \subseteq R^n \rightarrow R^m$ possessing a Fréchet derivative $A'(x)$ at $x \in int(D)$, for which the inverse $(A'(x))^{-1}$ exists. Then, for any sufficiently small $\varepsilon' > 0$, there is an $\varepsilon > 0$ such that, for a closed sphere $S = \bar{S}(x, \varepsilon)$,

$$\left| \sup_{x \neq x^{(1)} \in S} \frac{\|Ax^{(1)} - Ax\|}{\|x^{(1)} - x\|} - \|A'(x)\| \right| \leq \varepsilon' ,$$

$$\left| \inf_{x \neq x^{(2)} \in S} \frac{\|Ax^{(2)} - Ax\|}{\|x^{(2)} - x\|} - \|(A'(x)^{-1}\|^{-1} \right| \leq \varepsilon', \quad \text{and therefore}$$

$$\frac{\|A'(x)\| - \varepsilon'}{\|(A'(x))^{-1}\|^{-1} + \varepsilon'} \leq cond_1(A, S, x) \leq \frac{\|A'(x)\| + \varepsilon'}{\|(A'(x)^{-1}\|^{-1} - \varepsilon'} . \circ \qquad (2.55)$$

Hence, the problem (2.4) is well-posed at x if and only if the matrix condition number $cond_1(A'(x)) = \|A'(x)\| \cdot \|(A'(x))^{-1}\|$ is small. For the present problem, we can derive by formula (2.53) $cond_1(A'(x)) = cond_1(M(x))$. Therefore, the following theorem holds:

Theorem 2.55:

If the operator A is defined by formula (2.52), then the Fréchet derivative $A'(x)$ is singular whenever $x_1=0$, otherwise it is regular and an inverse matrix $(A'(x))^{-1}$ exists. In the latter case, we obtain

$$\frac{\|x\|}{|x_1|} \leq \operatorname{cond}_1(A'(x)) < \infty \;. \;\circ \tag{2.56}$$

*Proof:*

The determinant of a triangular matrix is equal to the product of its diagonal entries. Thus, $\det(M(x)) = x_1^n$ and $M(x)$ is singular iff $x_1=0$. If $x_1 \neq 0$, we choose the vectors

$v^{(1)} := (1,0,\ldots,0)^T$ and $v^{(2)} := (0,0,\ldots,0,1)^T$ in order to verify

$$\|M(x)\cdot v^{(1)}\| = \|x\| \leq \max_{v \neq 0} \frac{\|M(x)\cdot v\|}{\|v\|} = \|M(x)\| ,$$

$$\|M(x)\cdot v^{(2)}\| = |x_1| \geq \min_{v \neq 0} \frac{\|M(x)\cdot v\|}{\|v\|} = \|(M(x))^{-1}\|^{-1} \text{ . Hence,}$$

$\operatorname{cond}_1(M(x)) \geq \frac{\|x\|}{|x_1|}$ and the inequality (2.56) is valid. #

Based on an idea of FLEMMING a similarily structured upper bound, viz., $\operatorname{cond}_1(A'(x)) < n\cdot\left(\frac{\|x\|}{|x_1|}\right)^n$, can be found for the numerical condition, here.

## 2.2.5. Identification in Heat Equation Problems

An important class of identification and control problems is connected with the heat equation. Every linear or nonlinear problem of determining parameters or parameter functions that characterize initial temperatures, boundary temperatures and material functions arising in the differential equation belongs to this class. The observation data are generally obtained by measurements of temperature values at points that vary in space and in time. Sometimes, derived quantities as e.g. heat flux may also be measured. A survey of direct heat equation problems forming the basis for associated inverse problems is given by CARSLAW/JAEGER [58] and MIKHEYEV [301]. The calculus of heat problems also applies to diffusion (see e.g. CRANK [76], HÄFNER/GIESEL [185] and STOYAN [409-10]).

In this small section, we confine ourselves to identification problems for the one-dimensional heat equation. Let us begin with the linear heat equation problem backward in time (see ELDÉN [101], [104], FRANKLIN [131], GROETSCH [177, Chap.1], LATTES/LIONS [263], STRAKHOV [411], also [198] and [203]). We consider a metal bar of length one extending over the interval $0 \leq s \leq 1$. The heat conductivity and capacity of this solid are constituted so that the heat equation

$$\frac{\partial u(s,\tau)}{\partial \tau} = \frac{\partial^2 u(s,\tau)}{\partial s^2}, \quad 0 < s < 1, \quad 0 \leq \tau \leq 1, \tag{2.57}$$

holds for the temperature field $u(s,\tau)$. Here, $u(s,\tau)$ is a function varying with respect to the position $s$ of a considered point and with respect to the associated time $\tau$. At a zero time $\tau=0$ the bar had an initial temperature distribution

$$u(s,0) =: x(s), \quad 0 \leq s \leq 1 . \tag{2.58}$$

This initial function $x(s)$ is to be determined from measurements of final temperatures $b(s) := u(s,1)$, $0 \leq s \leq 1$, where zero Dirichlet data are imposed as follows:

$$u(0,\tau) = u(1,\tau) = 0, \quad 0 < \tau \leq 1 . \tag{2.59}$$

The seperation of variables yields

$$u(s,\tau) = \sum_{i=1}^{\infty} 2 \cdot \exp(-i^2\pi^2\tau) \cdot \left(\int_0^1 x(t) \cdot \sin i\pi t \cdot dt\right) \cdot \sin i\pi s . \tag{2.60}$$

From (2.60) it follows

$$b(s) = 2 \cdot \sum_{i=1}^{\infty} \exp(-i^2\pi^2) \cdot \left(\int_0^1 x(t) \cdot \sin i\pi t \cdot dt\right) \cdot \sin i\pi s$$ . The

relationship of the element $\mathfrak{x} = \langle x(t),\ 0 \leq t \leq 1 \rangle$ to the element $\mathfrak{b} = \langle b(s),\ 0 \leq s \leq 1 \rangle$ is expressed by the linear Fredholm integral equation of the first kind (2.44). If we consider the associated integral operator $\mathcal{A} : L_2[0,1] \to L_2[0,1]$, then we see that $\mathcal{A}$ is compact, injective and owing to the symmetry of its kernel selfadjoint. A complete singular system can be formed by

$\{\exp(-i^2\pi^2),\ \sqrt{2} \cdot \sin i\pi t,\ \sqrt{2} \cdot \sin i\pi t\}$, $i=1,2,\dots$ (see Lemma 2.36). Since the eigenvalues and continuous eigenfunctions of such an operator on $L_2[0,1]$ also represent eigenvalues and eigenfunctions of $\mathcal{A} : C[0,1] \to C[0,1]$, the backward heat equation problem is an instable ill-posed linear inverse problem whenever $B_1$ and $B_2$ coincide with $C[0,1]$ or $L_2[0,1]$ and $\mathcal{D}$ is a noncompact set of admissible solutions in $B_1$.

By means of two separate considerations, we are now going to be concerned with the cases $B_1 := B_2 := L_2[0,1]$ and $B_1 := B_2 := C[0,1]$, respectively. These brief studies will be aimed at finding additional restrictions, which can make the problem stable. In the former case $B_1 := B_2 := L_2[0,1]$, one can use the concept of logarithmic convexity (see e.g. [108]) in order to obtain continuous dependence on the data. For special situations, logarithmic convexity of a derived function allows us to show stability of a problem, although the additional restrictions are weaker than a compactness requirement on $\mathcal{D}$.

In the backward heat equation problem defined above, the function $f(\tau) := \int_0^1 u^2(s,\tau)\cdot ds$ is logarithmically convex, i.e., $\frac{d^2 \ln f(\tau)}{d\tau^2} \geq 0$.

Provided that the heat equation (2.57) with zero Dirichlet data (2.59) holds during the time interval $\tau_1 \leq \tau \leq \tau_2$, we have $\ln f(\lambda\cdot\tau_1 + (1-\lambda)\cdot\tau_2) \leq \lambda\cdot \ln f(\tau_1) + (1-\lambda)\cdot \ln f(\tau_2)$ if $0 \leq \lambda \leq 1$ and $\tau_1 < \tau_2$. A direct consequence of this property is the assertion of the following theorem (see [104] and [177]).

**Theorem 2.56:**

Consider on $B_1 = B_2 = L_2[0,1]$ the unconstrained linear identification problem $\mathcal{A}\varkappa = \ell$ associated with Eq. (2.44). If $\varkappa^{(1)} = \mathcal{A}^{\nu}\tilde{\varkappa}^{(1)}$ and $\varkappa^{(2)} = \mathcal{A}^{\nu}\tilde{\varkappa}^{(2)}$ holds for a certain number $\nu > 0$ and $\tilde{\varkappa}^{(1)}, \tilde{\varkappa}^{(2)} \in B_1$ satisfying $\|\tilde{\varkappa}^{(1)}\|_1 \leq \bar{c}$ and $\|\tilde{\varkappa}^{(2)}\|_1 \leq \bar{c}$, then

$$\|\varkappa^{(1)} - \varkappa^{(2)}\|_1 \leq 2\cdot \bar{c}^{\frac{1}{\nu+1}}\cdot \|\mathcal{A}\varkappa^{(1)} - \mathcal{A}\varkappa^{(2)}\|_2^{\frac{\nu}{\nu+1}} \; . \quad \circ \qquad (2.61)$$

**Remark 2.57:**

The physical meaning of such a so-called "source representation" $\varkappa = \mathcal{A}^{\nu}\tilde{\varkappa}$ ($\mathcal{A}^{\nu}$ - $\nu$-th power of $\mathcal{A}$) for the backward heat equation problem is the following: We imagine that the considered metal bar had a temperature distribution $\|u(.,-\nu)\|_{L_2[0,1]} \leq \bar{c}$ at a negative time $\tau_0 = -\nu$. In addition, we have to assume that the bar remains under conditions (2.57) and (2.59) during the time interval $-\tau_0 < \tau \leq 1$. $\circ$

Now let us take up the discussion of the latter case $B_1 := B_2 := C[0,1]$. In this context, we refer to the concept of uniform convergence in subintervals developed by TIKHONOV et al. [441, Chap.2] (see also MOROZOV et al. [158]). Here, the main properties of this approach will be mentioned without proof.

**Theorem 2.58:**

Consider an ill-posed linear or nonlinear identification problem (2.1) with $\mathcal{A} : \mathcal{D} \subseteq C[0,1] \to C[0,1]$ injective and satisfying the Assumption 2.1. Furthermore, consider a convergent subsequence $\{\ell^{(i)} = \mathcal{A}\varkappa^{(i)}\}_{i=1}^{\infty} \subseteq \mathcal{A}\mathcal{D}$ with $\|\ell^{(i)} - \ell\|_2 \to 0$ as $i \to \infty$ and $\ell = \mathcal{A}\varkappa \in \mathcal{A}\mathcal{D}$. We introduce a pair of closed convex subsets of $\mathcal{D}$ by the formulae

$$\tilde{\mathcal{D}} := \{\varkappa \in \mathcal{D} \subseteq C[0,1] : 0 \leq x(t_1) \leq x(t_2) \leq \bar{c} \quad \text{if} \quad 0 \leq t_1 < t_2 \leq 1\} \qquad (2.62)$$

(set of monotonically nondecreasing, nonnegative, uniformely bounded functions of $\mathcal{D}$ )

and

$$\tilde{\tilde{\mathfrak{D}}} := \left\{ \varkappa \in \mathfrak{D} \subseteq C[0,1] : \begin{array}{l} 0 \le x(t) \le \bar{c}\ ,\ x(\lambda \cdot t_1 + (1-\lambda) t_2) \le \lambda \cdot x(t_1) + (1-\lambda) x(t_2), \\ t_1 < t_2,\ 0 \le \lambda \le 1;\quad t, t_1, t_2 \in [0,1] \end{array} \right\} \tag{2.63}$$

(set of convex, nonnegative, uniformly bounded functions of $\mathfrak{D}$). If the functions $\varkappa^{(i)} = \langle x^{(i)}(t),\ 0 \le t \le 1 \rangle$, $i=1,2,\ldots$, and $\varkappa = \langle x(t),\ 0 \le t \le 1 \rangle$ belong either all to $\tilde{\mathfrak{D}}$ or all to $\tilde{\tilde{\mathfrak{D}}}$, then we have pointwise convergence

$$\lim_{i \to \infty} x^{(i)}(t) = x(t)\ ,\quad 0 < t < 1, \tag{2.64}$$

and, for any closed subinterval $[\varepsilon_1, 1-\varepsilon_2]$ with $0 < \varepsilon_1 < 1-\varepsilon_2 < 1$, uniform convergence

$$\lim_{i \to \infty} \| \varkappa^{(i)} - \varkappa \|_{C[\varepsilon_1, 1-\varepsilon_2]} = 0. \quad \circ \tag{2.65}$$

<u>Remark 2.59:</u>
With the exception of the boundary points $t=0$ and $t=1$, we get a continuous dependence of the solution to the backward heat equation problem upon the final temperature distribution at any point of the bar whenever we confine the admissible initial temperature functions to one of the sets $\tilde{\mathfrak{D}}$ or $\tilde{\tilde{\mathfrak{D}}}$. Note that neither $\tilde{\mathfrak{D}}$ nor $\tilde{\tilde{\mathfrak{D}}}$ is a compact subset of $C[0,1]$ if $\mathfrak{D}$ is noncompact. Thus, we can only derive that under the assumptions stated above a sequence $\{\varkappa^{(i)}\}_{i=1}^{\infty}$ as introduced in Theorem 2.58 has subsequences which are Cauchy sequences with respect to the $L_p[0,1]$-norm for arbitrary $1 \le p < \infty$ (see GONCHARSKI/JAGOLA [165]). This implies the limit conditions (2.64) and (2.65). $\circ$

<u>Remark 2.60:</u>
The proposition of Theorem 2.58 remains true if, in addition to the continuity, nonnegativity and boundedness, one requires that the considered functions are piecewise monotonous or alternating piecewise convex and concave over given subintervals $[t_i, t_{i+1}]$ , $i=1(1)k$, with $0=t_1<t_2< \ldots < t_{k-1}<t_k=1$ (see [156], [441]). $\circ$

Furthermore, note that a discretization using the projector $P^n$ of formulae (2.17), (2.18) provides sets of admissible skeleton solutions

$$\tilde{D} := P^n \tilde{\mathfrak{D}} = \{ x \in D \subseteq R^n :\ 0 \le x_i \le x_j \le \bar{c}\ ,\ 1 \le i < j \le n \} \tag{2.66}$$

and

$$\tilde{\tilde{D}} := P^n \tilde{\tilde{\mathfrak{D}}} = \{ x \in D \subseteq R^n :\ x_{i-1} - 2x_i + x_{i+1} \ge 0\ ,\ i=2(1)n-1;\ 0 \le x_i \le \bar{c},\ i=1(1)n \}, \tag{2.67}$$

which are polyhedronal compact closed convex subsets of $R^n$ (see [441, Sec. 2.3.]). Thus, a discretized version of the backward heat equation problem according to (2.17) through (2.19) is stable in the sense of Theorem 2.26 if D is a subset of one of the proposed sets $\tilde{D}$ or $\tilde{\tilde{D}}$. For further remarks on monotonicity and convexity constraints we also refer to FRIEDRICH et al. [138, Sec. 1.2.5.] and ECKHARDT [99].

The second identification problem corresponding to the heat equation, which we will briefly treat, is of nonlinear kind. On studying this problem let us consider a one-dimensional heat conduction or diffusion process in a nonhomogeneous medium so that the function of temperature or concentration $u(s,\tau)$ varying in space over $0 \leq s \leq 1$ and in time over $0 \leq \tau \leq 1$ satisfies the following initial-boundary value problem to the linear heat equation in cartesian coordinates (see STOYAN [408]):

$$\frac{\partial u(s,\tau)}{\partial \tau} = \frac{\partial}{\partial s}\left( a(s)\frac{\partial u(s,\tau)}{\partial s}\right) , \quad 0<s<1, \quad 0<\tau \leq 1 , \qquad (2.68)$$

$$u(s,0) = u^{(0)}(s) , \quad 0 \leq s \leq 1 , \qquad (2.69)$$

$$- a(s)\frac{\partial u(s,\tau)}{\partial s} = g(\tau) , \quad s=0 \text{ and } s=1 , \quad 0<\tau \leq 1 . \qquad (2.70)$$

We assume $u^{(0)}(s)$ and $g(\tau)$ to be sufficiently smooth given functions. On the other hand, $x(s) := a(s)$ represents the unknown spatially varying heat conductivity or diffusion coefficient function that we have to determine from temperature or concentration measurements $b(s) := u(s,1)$ at the final time 1. Provided that $B_1 := B_2 := C[0,1]$ and nonnegativity constraints are imposed, the operator $\mathcal{A} : \mathcal{D} \subseteq B_1 \rightarrow B_2$ transforming $\varkappa$ into $\ell$ is implicitly defined by (2.68) through (2.70). The operator $\mathcal{A}$ and therefore the inverse problem is nonlinear, although the differential equation (2.68) is a linear one. It will be assumed that the initial-boundary value problem presented above is uniquely solvable for any given $\varkappa \in \mathcal{D}$ so that the function $\mathcal{A}\varkappa$ is also uniquely determined.

A complete mathematical analysis of nonlinear identification problems in partial differntial equations is possible only for special cases (as to the theory and investigations on this topic see [49], [56], [57], [67], [233], [242-45], [353], [408] and [C23]). Uniqueness and stability of such an identification problem have been studied in detail for many particular cases (see e.g. GLASKO [151, Chap.3], ROMANOV [372], SEIDMAN [399], ISKENDEROV [220], DÜMMEL [96] and SUZUKI [417], [419]).

With the exception of some special well-behaved problems, nonlinear identification problems in the fields of heat conduction and heat transfer must be expected to be ill-posed. Consequently, their solutions may discontinuously depend upon the measurement data. For our example problem (2.68)-(2.70), STOYAN [408] has traced the instability and the corresponding ill-conditioning by numerical experiments (for spatially varying coefficients see also RICHTER [367]).

To find a discretized version (2.4) to an identification problem (2.1) with $\mathcal{A}$ given implicitly by an initial-boundary value problem, one can use the method of finite differences (see SAMARSKI [381], for inverse aspects, cf. e.g. [156] or [101]) or the finite element method. The associated operator A, in general, will also be given implicitly, e.g. by a difference scheme. For nonlinear operators $\mathcal{A}$, it frequently seems to be natural to choose A nonlinear, too. As we know, the error of discretization $\|Q^m\mathcal{A}\varkappa - AP^n\varkappa\|$ corresponds to the choice of $P^n$ and $Q^m$. Besides, this deviation also depends on numerical control parameters like stepsizes in space and in time and on stopping criteria for iterative refinement techniques. All these control parameters are closely related to the used difference scheme forming A. The mentioned numerical auxiliary values are fairly independent of the chosen dimensions m and n. A coarsely meshed difference scheme for A generates an essential additional discretization error. However, given x, the expense of computing a vector Ax (total amount of computational work for solving the direct problem) considerably grows if the difference scheme stepsizes (especially time stepsizes in parabolic problems) are decreased substantially. Since solving an identification problem associated with differential equations always requires a many times solution of the direct problem, this economical point of view is if great importance for the practical solution of inverse heat equation problems.

At the end of this small report on inverse heat equation problems we are going to refer to the most important case of identification of material functions in the quasilinear parabolic equation. Let us impose polar coordinates and boundary conditions of the third kind. Thus, we consider the initial-boundary value problem

$$c(u)\cdot\frac{\partial u(r,\tau)}{\partial\tau} = \frac{1}{r}\cdot\frac{\partial}{\partial r}\cdot\left(r\cdot a(u)\cdot\frac{\partial u(r,\tau)}{\partial r}\right), \quad 0<r<1, \quad 0<\tau\leq 1, \tag{2.71}$$

$$u(r,0) = u^{(0)}(r), \quad 0\leq r\leq 1, \tag{2.72}$$

$$\frac{\partial u(r,\tau)}{\partial r} = 0, \quad r=0, \quad 0<\tau\leq 1, \tag{2.73}$$

$$-a(u)\frac{\partial u(r,\tau)}{\partial r} = \gamma(u)\cdot(u(r,\tau) - \Psi(\tau)), \quad r=1, \quad 0<\tau\leq 1. \tag{2.74}$$

The problem (2.71) through (2.74) arises if a long, symmetric metal rod possessing radius one and initial temperature $u^{(0)}(r)$ at time zero is heated by a gas stream. The gas stream temperature $\Psi(t)$ and the initial profile $u^{(0)}(r)$ are assumed to be known and their function values have to lie in the temperature interval $[0,U]$. By the maximum principle (see MIKHLIN [302]) this implies $0 \leq u(r,\tau) \leq U$, $0 \leq r \leq 1$, $0 \leq \tau \leq 1$ for the temperature behaviour.

Indeed, for applied problems, the temperature dependence of the material functions $c(u)$ (density multiplied by heat capacity), $a(u)$ (heat conductivity) and of the process function $\gamma(u)$ (heat transfer coefficient) must not be neglected. For the present problem, identification denotes the determination of these three functions defined on the interval $[0,U]$ from temperature measurements or the determination of one or two functions of this triple if the remaining functions are already known.

In any case, the associated operator $\mathcal{A}$ is nonlinear and so is the inverse problem. Note that Eq. (2.71) embodies a nonlinear parabolic differential equation. Many authors have studied nonlinear inverse problems of a structure similar to (2.71)-(2.74) that correspond to the quasilinear heat equation (see CANNON/DU CHATEAU [50-1] , ERIKSSON/DAHLQUIST [119] , GLASKO et al. [152], GOLDMAN [156], HOFFMANN [193], [195], KRUKOVSKI [254], also [185], [205], [207], [213-4], [275]). Here, we renounce repeating all the results presented in the papers mentioned above. However, we are going to make some remarks on the associated discretised versions. In addition to the above general notes on using difference schemes for discretized parabolic problems, it should be mentioned that the nonlinear character of Eqs. (2.71) and (2.74) requires iterative schemes in order to solve the nonlinear direct problem sufficiently well (see e.g. [297] or [481]).

The operator $P^n$ of a priori discretization can be chosen according to (2.17), (2.18) transformed to the interval $[0,U]$ if an unknown function $x(u)$ (e.g. $a(u)$, $c(u)$ or $\gamma(u)$) is submitted and a sufficiently large dimension number n gets possible (see [207]). Another frequently occurring situation is the following (see [205], [235], [260] or [368]): Let $\langle \varphi^{(i)}(u),\ 0 \leq u \leq U \rangle$, $i=1,2,\ldots$, be a basis in $B_1 := C[0,U]$. Then

$$P^n \varkappa := x = (x_1,\ldots,x_n)^T \in R^n \quad \text{if } x(u) = \sum_{i=1}^{\infty} x_i \cdot \varphi^{(i)}(u)\ ,\ 0 \leq u \leq U, \quad (2.75)$$

is a suitable projector whenever we know from physical laws that the unknown function $x(t)$ at least approximately belongs to the n-dimensional subspace span($\varphi^{(1)},\ldots,\varphi^{(n)}$), where n should generally be small. For instance, the heat conductivity function a(u) is often a good approximation to a linear function

$$a(u) := x_1 + x_2 \cdot u \; . \qquad (2.76)$$

On the other hand, the transfer function $\gamma(u)$ may frequently be represented by

$$\gamma(u) := x_1 + \frac{x_2}{u} \qquad \text{or} \qquad \gamma(u) := x_1 + x_2 \cdot u + x_3 \cdot u^2 \qquad (2.77)$$

in order to express a water-metal heat transfer or the Stefan-Boltzmann law of heat transfer, respectively.

If we assume the temperature field $\mathcal{b} = \langle u(r,\tau),\ 0 \leq r \leq 1,\ 0 \leq \tau \leq 1 \rangle$ to be a continuous function with respect to both variables, i.e., $B_2 := C([0,1]\times[0,1])$, then the projector $Q^m$ will be defined by

$$Q^m \mathcal{b} := b = (b_1,\ldots,b_m)^T;\ b_i = u(p_i),\ p_i = (s_i,\tau_i)^T \in [0,1]\times[0,1] \; . \qquad (2.78)$$

Thus, $Q^m$, the operator of experimental design (measurement planning), is determined by choosing an m-tuple of distinct points $p_i$ from the considered radius-time rectangle $[0,1]\times[0,1]$. There are very different experimental design strategies as follows:

a. The time varying measurement strategy -
   $s_i$= const., $\tau_i$ distinct (one sensor, m observed time points).

b. The spatially varying measurement strategy -
   $s_i$ distinct, $\tau_i$= const. ( m sensors, one observation time).

c. Mixed strategies -
   neither $s_i$ nor $\tau_i$ constant.

Many authors have searched for an optimal experimental design: find a maximum of information about the unknown material and process functions with a minimum expense for the measurements. It is required that the arising discretized identification problems (2.4) are as best as possible well-conditioned. Then, the inevitable observation and discretization errors do not falsify the solutions considerably. For this subject, we refer to detailed studies published in [C23, Sec. 2.4.] and [205].

In the former paper [C23], the scalar heat conductivity and heat transfer coefficients are to be determined simultaneously during the heating of a metal cube. The behaviour of the numerical condition number may be studied when the measurements of temperature vary in space and in time. The more the influence of one of the phenomena heat conduction and heat transfer is predominant for a measurement point, the more information about the associated coefficient the corresponding observation data contain. If m=2 data are used in order to reconstruct the n=2 unknowns heat condictivity and heat transfer coefficient, then the associated two dimensional nonlinear system Ax=b is well-conditioned if one sensor has been located at the mid-point of the cube and another sensor near a corner. If only one sensor is available for both data, the difference of measurement times should be sufficiently large.

In the latter article, the identification of heat conductivity a(u) according to (2.71)-(2.74) and (2.76) is investigated with respect to different measurement strategies. If one considers the sensitivity of the solution in dependence of perturbations in the data (and thus implicitly the numerical condition behaviour), then the time varying measurement strategy a. yields more information about a(u) than the spatially varying strategy b. provided that m data are taken into consideration. Frequently, the components of $Q^m \mathcal{A}\varkappa - AP^n \varkappa$ tend to be uniformly positive or uniformly negative whenever implicit difference schemes generate A. This feature may imply some surprising effects. Thus, in [205] a numerical example is presented, which shows that a growing number m of data may lead to a growing reconstruction error. As a rule, the reconstruction error tends to decrease as m grows and therefore the degree of information of the data is improved. However, this is only true if the discretization error can approximately be considered as a random vector with zero mean (see Chap. 5). If the discretization error components are sign resistant, then an additional systematic error in the solution has to be expected. This additional error may grow with m sometimes.

### 2.2.6. Identification in a Coupled System of Integral Equations of the First Kind

The electric behaviour of thin layers of semi-conductor material is essentially determined by its conductivity. This property can be derived from a pair of physical quantities: the concentration $\eta(t)$ and the mobility $\mu(t)$ of charge carriers in the layer as a spatially varying function over a scale $0 \leq t \leq 1$, where t is the perpendicular distance from the bottom of the layer. The functions $\eta(t)$ and $\mu(t)$ are not directly observable. However, effective values (derived quantities) $\nu_{eff}^{(1)}(s)$ and $\nu_{eff}^{(2)}(s)$ can be measured if a gate voltage $s \in [0, S]$ is applied.

The problem of identifying $\varkappa = \langle (\eta(t), \mu(t)) \in C[0,1] \times C[0,1] =: B_1 \rangle$

from $\mathcal{b} = \langle (v_{eff}^{(1)}(s),\ v_{eff}^{(2)}(s)) \in C[0,S] \times C[0,S] =: B_2 \rangle$ is, written in mathematical terms, a coupled system of two linear integral equations of the first kind (see also [190]) with given differentiable kernel function $k(s,t)$:

$$\begin{aligned} v_{eff}^{(1)}(s) &= \int_0^1 k(s,t)\cdot\eta(t)\cdot\mu(t)\cdot dt\ , \\ & \qquad\qquad\qquad\qquad\qquad k(s,t) \geq \underline{k} > 0, \\ v_{eff}^{(2)}(s) &= \int_0^1 k(s,t)\cdot\eta(t)\cdot\mu^2(t)\cdot dt\ , \quad 0 \leq t \leq 1\ ,\ 0 \leq s \leq S\ . \end{aligned} \tag{2.79}$$

Remark 2.61:

To find the composite functions $\eta(t)\cdot\mu(t)$ and $\eta(t)\cdot\mu^2(t)$, respectively, we have to solve a Fredholm integral equation of the first kind (see Sec. 2.2.1.). For instance, the function $\mu(t)$ could be calculated as a quotient of the solutions to a couple of linear instable identification problems. However, the solution strategy of decoupling the equations in (2.79) is useless, since the particular solution errors to both equations may be superposed in such an unfavourable way by division that the approximations to $\varkappa$ thus obtained are extremely bad and physically noninterpretable. ○

On the other hand, we can solve the coupled system as a nonlinear (and as one may easily derive instable ill-posed) identification problem (2.1). The associated continuous operator $\mathcal{A}: \mathcal{D} \subseteq B_1 \to B_2$ with $\mathcal{D}$ prescribing nonnegativity constraints for $\eta(t)$ and $\mu(t)$, $0 \leq t \leq 1$, is even defined explicitly by the system (2.79). Again, this enables an explicit representation of a discretized version (2.4) to (2.79). We are now going to introduce one of the possible discrete fashions.

Let us proceed with exactly the same construction as in the discretization (2.17)-(2.19), but here applied to both equations of (2.79), simultaneously:

$$\begin{aligned} t_i &:= \left(\frac{2i-1}{2}\right)\cdot h\ ,\ h := \frac{1}{n_o}\ ,\ n := 2n_o,\ i=1(1)n_o; \\ s_j &:= j\cdot\tau\ ,\ \tau := \frac{S}{m_o}\ ,\ m := 2m_o\ ,\ j=1(1)m_o\ ; \end{aligned} \tag{2.80}$$

$$\begin{aligned} P^n\varkappa := x &= \begin{bmatrix} P_1^{n_o}\varkappa \\ P_2^{n_o}\varkappa \end{bmatrix} = (\eta(t_1),\dots,\eta(t_{n_o}),\mu(t_1),\dots,\mu(t_{n_o}))^T, \\ Q^m\mathcal{b} := b &= \begin{bmatrix} Q_1^{m_o}\mathcal{b} \\ Q_2^{m_o}\mathcal{b} \end{bmatrix} = (v_{eff}^{(1)}(s_1),\dots,v_{eff}^{(1)}(s_{m_o}),v_{eff}^{(2)}(s_1),\dots,v_{eff}^{(2)}(s_{m_o}))^T \end{aligned} \tag{2.81}$$

and, for $A: D \subseteq R^{2n_o} \to R^{2m_o}$, $D = \{x \in R^n: x_i \geq 0, i=1(1)n\}$,

$$(Ax)_j = \begin{cases} \sum_{i=1}^{n_o} h \cdot k(s_j, t_i) \cdot x_i \cdot x_{i+n_o}, & 1 \leq j \leq m_o \\ \sum_{i=1}^{n_o} h \cdot k(s_{j-m_o}, t_i) \cdot x_i \cdot x_{i+n_o}^2, & m_o+1 \leq j \leq m \end{cases} . \quad (2.82)$$

Therefore, $Ax=b$ represents a system of $2m_o$ nonlinear equations in $2n_o$ unknowns. From formula (2.82) we obtain a Fréchet derivative $A'(x) \in R^{m \times n}$ in block-matrix form at a point $x \in D$ as follows:

$$\begin{aligned} A'(x) &= \begin{pmatrix} M_1(x) & M_2(x) \\ M_3(x) & M_4(x) \end{pmatrix}, \quad M_\nu(x) \in R^{m_o \times n_o}, \nu=1(1)4, \\ M_1(x) &= ( h \cdot k(s_j, t_i) \cdot x_{i+n_o} )_{i=1(1)n_o}^{j=1(1)m_o}, \\ M_2(x) &= ( h \cdot k(s_j, t_i) \cdot x_i )_{i=1(1)n_o}^{j=1(1)m_o}, \\ M_3(x) &= ( h \cdot k(s_j, t_i) \cdot x_{i+n_o}^2 )_{i=1(1)n_o}^{j=1(1)m_o}, \\ M_4(x) &= ( 2h \cdot k(s_j, t_i) \cdot x_i \cdot x_{i+n_o} )_{i=1(1)n_o}^{j=1(1)m_o} . \end{aligned} \quad (2.83)$$

We can generalize the idea of Theorem 2.54 to the case of a rectangular matrix $A'(x)$ if considering formula (2.6). Then the numerical condition number $cond_1(.)$ of a nonlinear identification problem $Ax=b$, $A: R^n \to R^m$, $m \neq n$ at a point $x \in R^n$ will be determined by the matrix condition number $cond_1(A'(x)) = \frac{\sigma_1}{\sigma_n}$ of $A'(x)$. If $A'(x)$ fails to be a full-rank matrix with $rank(A)=n$, then the condition number becomes infinite. In formula (2.83), this occurs if $x_i=0$ for a certain integer i, since in such a case a whole column of $A'(x)$ vanishes. For continuity reasons, the problem (2.4) according to (2.80)-(2.82) gets more and more ill-conditioned as at least one component $x_i>0$ tends to zero.

On the other hand, if $\varkappa = \langle (\eta(t), \mu(t)) \in C[0,1] \times C[0,1], \eta(t) \geq \underline{c} > 0, \mu(t) \geq \underline{c} > 0, 0 \leq t \leq 1 \rangle$ and $x = P^n \varkappa$, then any block $M_\nu(x)$, $\nu=1(1)4$ of $A'(x)$ can be interpreted as a discretization matrix (2.19) of a Fredholm integral operator of the first kind with smooth composite kernels $k(s,t) \cdot \mu(t)$, $k(s,t) \cdot \eta(t)$, $k(s,t) \cdot \mu^2(t)$ and $2k(s,t) \cdot \eta(t) \cdot \mu(t)$. Therefore, the condition behaviour of the problem is expressed by Theorem 2.32. The finer the discretization ($h \to 0$), the larger the associated condition number becomes.

## 2.2.7. On a Parabolic Boundary Control Problem Arising in Rheology

The presented series of inverse problems will be closed by two examples of nonlinear control problems (cf. formulae (2.3) and (2.5)). We shall not discuss linear control problems in detail, since these problems are intimately related to linear Fredholm or Volterra equations of the first kind. However, the main properties of those classes of inverse problems are summarized in Secs. 2.2.1. and 2.2.2., respectively. As we know, identification and control problems are different in their intrinsic behaviour with respect to the uniqueness requirement. For linear control problems, the assumption $\mathbb{N}(\mathcal{A}) = \{0\}$ can be omitted. On the other hand, the ill-conditioning of discretized versions to ill-posed linear problems, for a sufficiently large number n established by Theorems 2.32 and 2.43, generally does not disappear in the case of discretized linear control problems.

There exists a special kind of smooth kernel functions $k(s,t)$ in Fredholm integral equations such that the corresponding discrete control problems are very well-conditioned, although the associated discretized identification problems are ill-conditioned. Namely, if $k(s,t) \equiv k_0(s) > 0$, for all considered t, then a matrix A won by formula (2.19) is of rank one. All the entries of a fixed row of A have the same positive value. Hence, $\mathrm{cond}_1(A,S,x) = \infty$ and $\mathrm{cond}_2(A) := \|A\| \cdot \|A^+\| = 1$. Kernel functions of this fashion are very unfavourable for identification, since the right-hand side $\mathcal{b}$ contains almost no information about $\varkappa$. However, elements $\varkappa$ of a wide set are compatible with a desired element $\mathcal{b}$ with respect to a sufficiently small residual norm $\|\mathcal{A}\varkappa - \mathcal{b}\|_2$. The numerical condition behaviour discussed above is the discrete analogue to this phenomenon.

Now let us pick up a control problem from rheology (see SCHNEIDER [384-5], also [209, Appendix]). We imagine a homogeneous isotropic viscous fluid (e.g. a melt of glass or polymer) with properties which are dependent on temperature and pressure. This fluid is assumed to flow through a vertical straight circular tube of length one. Moreover, the fluid adheres to the wall and the heat transfer between fluid and wall will be described, in mathematical terms, by boundary conditions of the first kind. The variables $v_r(r,l)$, $v_l(r,l)$, $p(l)$ and $u(r,l)$ denote the quantities radial component of the fluid velocity, axial component of the fluid velocity, pressure and temperature, respectively. These functions are spatially varying with respect to the radial coordinate $0 \leq r \leq 1$ ($r=0$ - centre of the tube, $r=1$ - wall of the tube) and with respect to the axial coordinate $0 \leq l \leq 1$.

Then, the direct problem is determined by an initial-boundary value problem to a system of partial differential equations

$$\frac{dp}{dl} = \frac{1}{c_1} \cdot \frac{1}{r} \cdot \frac{\partial}{\partial r}\left(r \cdot \eta \cdot \frac{\partial v_l}{\partial r}\right), \quad 0 \leq l \leq 1, \quad 0 < r < 1, \tag{2.84}$$

$$\int_0^1 \varrho \cdot v_l \cdot r \cdot dr = \frac{1}{2}, \quad 0 \leq l \leq 1, \tag{2.85}$$

$$v_l \frac{\partial \varrho}{\partial l} + \varrho\left(\frac{\partial v_r}{\partial r} + \frac{v_r}{r} + \frac{\partial v_l}{\partial l}\right) = 0, \quad 0 \leq l \leq 1, \quad 0 < r < 1, \tag{2.86}$$

$$\varrho \cdot c \cdot c_1 \cdot c_2 \cdot \left(v_r \frac{\partial u}{\partial r} + v_l \frac{\partial u}{\partial l}\right) = \frac{1}{r} \cdot \frac{\partial}{\partial r}\left(r \cdot \lambda \cdot \frac{\partial u}{\partial r}\right), \quad \begin{matrix} 0 \leq l \leq 1, \\ 0 < r < 1, \end{matrix} \tag{2.87}$$

$$p(0) = p_0, \; u(r,0) = u^{(0)}(r), \; v_r(r,0) = 0, \; 0 \leq r \leq 1, \tag{2.88}$$

$$v_r(1,l) = v_l(1,l) = 0, \quad 0 \leq l \leq 1, \tag{2.89}$$

$$u(1,l) =: x(l), \quad 0 \leq l \leq 1, \tag{2.90}$$

with $\eta$ the viscosity, $\lambda$ the heat conductivity, $\varrho$ the density, c the heat capacity and $c_1$, $c_2$ other constants of rheology. All material functions and constants are assumed to be known. We denote by $\mathcal{A}x := \langle u(r,1),\ 0 \leq r \leq 1 \rangle$ the radial temperature profile when the fluid has passed the tube.

The control problem is aimed at finding a boundary temperature function $x(l)$, $0 \leq l \leq 1$, along the tube wall so that a desired final temperature profile $b(r) := u(r,1)$, $0 \leq r \leq 1$, is approximated as best as possible. This is a problem (2.3) with $\mathcal{A}: \mathcal{D} \subseteq B_1 := C[0,1]$ representing a nonlinear operator that transformes the boundary control $x(l)$ into the final profile $u(r,1)$. The operator $\mathcal{A}$ is implicitly defined by (2.84) through (2.90), where the parabolic equation (2.87) substantially influences the character of the system. Technical restrictions suggest choosing the closed convex domain $\mathcal{D} = \{x(l) \in C[0,1]: 0 \leq \underline{x}(l) \leq x(l) \leq \overline{x}(l),\ 0 \leq l \leq 1\}$, where $\underline{x}(l)$ is a convex continuous function and $\overline{x}(l)$ a concave continuous function over $[0,1]$.

The analytic structure of the present nonlinear control problem is rather complicated. Therefore, an extensive stability analysis seems to be hard to perform. However, numerical experiments show the ill-posedness of the problem. Many authors have dealt with parabolic boundary control problems and have mentioned ill-posedness effects (see e.g. SEIDMAN [396], [399]). For similar control problems, in general, compare LIONS [280] , CHAVENT [64], MACKENROTH [289] and TRÖLTZSCH [445-7].

Referring back to our rheological problem, a projector $P^n$: $C[0,1]\to R^n$ is given in a natural way by the technical equipment of the tube. If n is the number of equally spaced heating elements along the tube wall, the formulae (2.17) and (2.18) can be applied with respect to $P^n$, where $t:=l$, $\underline{t}:=0$, $\overline{t}:=1$, $t_i:=l_i$. An iterative difference method discussed in [384] allows us to construct a nonlinear operator $A: D\subseteq R^n\to R^m$ with $D=\{x\in R^n: \underline{x}(l_i)\leq x_i\leq \overline{x}(l_i),\ i=1(1)n\}$. The operator $Q^m$: $C[0,1]\to R^m$ is chosen by the procedure $Q^m \mathscr{b} := b=(b(r_1),\ldots, b(r_m))^T$, where the grid $r_j$, $j=1(1)m$, over the radius interval corresponds to the used radial discretization according to the finite difference scheme generating A.

In [209, Appendix], results of a numerical experiment on the present problem with n=6 and m=50 are reported. Fairly large condition numbers may occur even if n is very small. As an interesting fact the study has pointed out that the nonlinear functional $\|Ax-b\|^2$ to be minimized subject to given upper and lower control limitations possesses a few different local minima, for which the associated functional values are almost equal to $\min_{x\in D}\|Ax-b\|^2$.

Control vectors $x\in D$ with a very different monotonicity behaviour may lead to almost optimal residual norms. Bang-bang properties, as they arise in linear parabolic equation control problems (see e.g. [150], [446-71]), are completely missing here. A sensitivity analysis indicates that small perturbations of the control x(l) cause significant changes in the residual norms $\|\mathscr{A}\mathscr{x}-\mathscr{b}\|_2$ and $\|Ax-b\|$ if and only if $l\approx 1$, i.e., if l denotes a point close to the bottom of the vertical tube. However, changes of x(l) with $l\approx 0$ may also influence the final temperature profile u(r,1) whenever they are large enough.

### 2.2.8. An Inverse Eigenvalue Problem

The final example, which we consider now, differs from those of the previous sections with respect to the nature of the spaces $B_1$ and $B_2$. For the present problem, we have $B_1:=R^n$ and $B_2:=R^m$. Therefore, continuous and discrete version of the problem do not have to be distinguished. Let $M_i\in R^{k\times k}$, $i=1(1)n$, be a given finite sequence of real symmetric positive semidefinite matrices. Moreover, consider the derived matrix M(x) defined by the formula

$$M(x):=x_1M_1+x_2M_2+\ldots+x_nM_n\ ,\quad x\in D=\{x'\in R^n:\ x_i'\geq 0,\ i=1(1)n\}\ ,\qquad (2.91)$$

which possesses the same properties as $M_i$. Furthermore, for any vector $x\in D$, let us have a look at the special matrix eigenvalue problem

$$M(x)\cdot v^{(i)}(x) = \lambda_i(x)\cdot v^{(i)}(x)\ ,\quad \lambda_i(x)\geq 0,\ 0\neq v^{(i)}(x)\in R^k,\ i=1(1)n, \tag{2.92}$$

where $\lambda_1(x)\geq\lambda_2(x)\geq\ldots\geq\lambda_k(x)\geq 0$ is the uniquely determined nonincreasing sequence of eigenvalues of $M(x)$ and $\{v^{(i)}(x)\}_{i=1}^k$ an associated orthonormal system of eigenvectors.

We will briefly treat a control problem from the family of inverse eigenvalue problems (for the theory of this class of inverse problems see KUHNERT [256-7], also [25] and [405]): Given a vector $b\in R^m$ with $m\leq k$ containing m desired eigenvalues, we search for a vector $x\in D$ so that the m largest eigenvalues of $M(x)$ approximate the components $b_1\geq b_2\geq\ldots\geq b_m$ of the vector b as best as possible. That means, the sum of squares $\sum_{j=1}^m(\lambda_j(x)-b_j)^2$ is to be minimized. We thus obtain a problem of the form (2.5), where the operator $A\colon D\subseteq R^n\to R^m$ transforming x into $(\lambda_1(x),\ldots,\lambda_m(x))^T$ acts according to (2.91) and (2.92).

Note that the inverse eigenvalue problem under consideration arises in mechanics if a vibrating system with n degrees of freedom is controlled. In such a case, x denotes the vector of mass parameters and the matrices $M_i$ express the stiffness of the system. The control problem is aimed at avoiding undesirable resonance effects by prescribing eigenvalues $b_j$, $j=1(1)m$, which are well-separated from critical eigenvalues.

Remark 2.62:

The control problem introduced above is, in general, a nonlinear one. Only in the particular case $M_i := v^{(i)}\cdot v^{(i)T}$, $(v^{(i)},v^{(j)})=0$ if $i=1(1)n$, $j=1(1)n$, $i\neq j$, the eigenvalues $\lambda_j$, $j=1(1)k$ linearly depend on x. The continuity of A directly results from the Poincaré-Fischer extremum principle (see Lemma 2.44) applied to the matrix (2.91). The problem may be instable and ill-posed (see e.g. [257, Example 2]). It tends to become ill-conditioned at least if the number n of control components increases. ○

Supposed that $x\in D$ is chosen so that all the associated eigenvalues $\lambda_j(x)$, $j=1(1)m$ are distinct, then the Fréchet derivative $A'(x)$ is given by the formula (see [405])

$$A'(x) = ((M_i v^{(j)}(x),\ v^{(j)}(x))_{i=1(1)n}^{j=1(1)m}\ . \tag{2.93}$$

Then we can use the quotient $\frac{\sigma_1}{\sigma_r}$ of the singular values of $A'(x)$ with $r:=\mathrm{rank}(A'(x))$ in order to evaluate the numerical condition of the control problem. From formula (2.93) it follows that the well-conditioning or ill-conditioning of the problem is caused by two different features.

On the one hand, the more the matrices $M_i$, $i=1(1)n$, are almost linearly dependent, the more the columns of $A'(x)$ are also approximately linearly dependent. However, if one of the matrices is exactly a linear combination of the remaining matrices, i.e., $M_\nu \in \mathrm{span}(M_1,\ldots,M_{\nu-1},M_{\nu+1},\ldots,M_n)$, this does not enlarge the condition number. Here, again the particularity of control problems compared to identification problems is refelcted.

On the other hand, if the difference $|\lambda_j(x)-\lambda_{j+1}(x)|$ between two eigenvalues gets small, then two neighbouring rows of $A'(x)$ may become almost linearly dependent. At least if $m \leq n$, this implies ill-conditioning of the control problem. The former case of ill-conditioning is independent of of the actually used $x \in D$, whereas the latter case appears for specific vectors $x$ only.

## 2.3. On A Priori Information

By the examples of Sec. 2.2. we have seen that ill-posedness is characteristic of inverse problems (2.1) and (2.3) as well as ill-conditioning is characteristic of the discretized versions (2.4) and (2.5). Perturbations of the observation data, even if they are very small, may lead to significant errors in the solutions. In order to overcome the instability and nonuniqueness of the solution to inverse problems, it seems to be very important to use any available information to the full extent. There are two classes of a priori information about the solution $\mathcal{X}$ of an inverse problem or about its skeleton $x$, respectively. We have to distinguish objective and subjective a priori information. If any law of nature or any technical confinement prescribes constraints for the element to be identified or for the control function to be found, then these prescriptions provide objective a priori information. The adjective "objective" is due to the high degree of reliability of this kind of information. In our mathematical models, objective a priori information is predominantly expressed by the domains $\mathcal{D}$ and D (sets of admissible solutions).

The other class of subjective a priori information embodies a lower degree of reliability. For instance, many functions arising in the nature are smooth. Therefore, one can expect the solution we search for to be smooth at least over subintervals. This is a subjective information, which can be expressed, in mathematical terms, by introducing a nonnegative functional that evaluates the smoothness of a potential solution. If $\varkappa = \langle x(t),\ 0 \leq t \leq 1 \rangle$ is a differentiable function with a quadratically integrable first derivative $\varkappa' = \langle x'(t),\ 0 \leq t \leq 1 \rangle$, then

$$\Pi(\varkappa) = p_0 \cdot \int_0^1 x^2(t)dt + p_1 \cdot \int_0^1 (x'(t))^2 dt = p_0 \cdot \|\varkappa\|^2_{L_2[0,1]} + p_1 \cdot \|\varkappa'\|^2_{L_2[0,1]} \tag{2.94}$$

with nonnegative weights $p_0$ and $p_1$ is a frequently used smoothing functional and

$$\Omega(x) = \tilde{p}_0 \cdot \sum_{i=1}^{n} x_i^2 + \tilde{p}_1 \cdot \sum_{i=1}^{n-1} (x_{i+1} - x_i)^2 \ , \quad \tilde{p}_0, \tilde{p}_1 \geq 0 \ , \tag{2.95}$$

its discrete analogue. Moreover, the weighted $L_2$-norms of higher derivatives $x''(t)$, $x'''(t)$ and so on (in the classical sense or in the generalized sense of Sobolev spaces) may be added to the above formulae. The maximum occurring order of derivatives in the smoothing functionals and the weights $p_0, p_1, \ldots$ have to be chosen subjectively with respect to the expected smoothness of the solution (cf. CULLUM [79]).

Besides, the evaluation functionals can also result from other subjective information. They may involve relations between the values of the function $\varkappa$ for different arguments. Moreover, probability distributions over the domain of admissible solutions are sometimes taken into account. If we summarize the above remarks on subjective a priori information, it can be stated that we have to establish a nonnegative evaluation functional that takes on small values if and only if the associated argument element $\varkappa$ or the argument vector $x$ is suspected to be a solution to the inverse problem.

Within Chapter 3, mathematical requirements for smoothing and other evaluation functionals are formulated. In addition to objective and subjective a priori information we have to distinguish a further pair of classes of information: deterministic and stochastic. Consider these two kinds of a priori information with respect to the character of the data and observation errors in discretized identification and control problems (2.4) and (2.5).

We assume the vector

$$z := b + y = Q^m \ell + y \in R^m \tag{2.96}$$

to be the complete set of available real data in solving the discretized inverse problem by a digital computer. Imagine that the exact right-hand side vector $b$ is superposed by an error vector $y \in R^m$. In the identification case, this will generally be a measurement error vector with components of significant magnitude. For solving control problems numerically, the vector b of wanted values is to be prepared for the digital computer. Then, y may be reduced to the roundoff error, however, it is never entirely missing.

<u>Definition 2.63:</u>
In the sequel, the problem of finding $x \in D$ from $z \in R^m$ according to (2.4), (2.96) or (2.5), (2.96) is called a noisy data discretized inverse problem (identification or control problem, respectively). For short, we will sometimes say noisy data problem. Furthermore, let us distinguish deterministic and stochastic noisy data problems in dependence of the deterministic or stochastic character of the error vector y. ○

<u>Assumption 2.64:</u>
We assume that, for a deterministic noisy data problem, there is an error bound $\delta > 0$ such that

$$\|y\| \leq \delta . \tag{2.97}$$

If considering a stochastic noisy data problem, then assume that the really arising error vector y is a realization of a centralized random vector $\eta$ with zero mean and given symmetric positive definite covariance matrix $C \in R^{m \times m}$. In addition, we assume to know the kind of probability distribution. Frequently, Gaussian distribution

$$\eta \sim \mathcal{N}(0,C), \quad 0 \in R^m, \; C \in R^{m \times m} \tag{2.98}$$

will be required. ○

If it is justified to randomize the space of solutions, then the associated probability distribution parameters may be used to solve noisy data problems.

<u>Definition 2.65:</u>
We say that a stochastic noisy data problem is Bayesian if any skeleton solution vector $x \in R^n$ is considered to be a realization of a random vector $\xi$ with given probability distribution, mean vector $\bar{x} \in R^n$ and a symmetric positive definite covariance matrix $B \in R^{n \times n}$. Frequently, Gaussian distribution

$$\xi \sim \mathcal{N}(\bar{x},B), \quad \bar{x} \in R^n, \; B \in R^{n \times n} \tag{2.99}$$

will be required. If the probability distribution of $\xi$ or at least the moments $\bar{x}$ and B are not known, however, a sample $x^{(1)},\dots,x^{(k)}$ of other realizations of $\xi$ is available, then we call the problem empirical Bayesian noisy data problem. If the solution space is not randomizable, then the problem is termed non-Bayesian. ○

It seems to be clear that a stochastic noisy data problem, in general, is more informative than the corresponding deterministic counterpart. Namely, a whole probability distribution much more reveals the specific behaviour of the input data errors compared to a single upper bound value $\delta$ in the deterministic case. This idea also applies to the difference between Bayesian and non-Bayesian noisy data problems. However, the stochastic treatment of noisy data problems in any case requires a large set of additional data, e.g. $\frac{n(n+3)}{2}$ real numbers in order to fill in mean and covariance matrix of an n-dimensional random vector. Seldom all these data are immediately available. On the other hand, the covariance matrix C may be calculated by a measurement tool analysis.

Randomization is reasonable only for specific applied problems with physical or technical background, which justifies the Bayesian approach. If the problem is randomizable, then the empirical Bayesian strategy often represents an appropriate way to find the required distribution parameters by an estimation based on sample vectors. Note that, for stochastic noisy data problems, the reconstruction of x from z is an estimation problem. In the empirical Bayesian case, we have a two-staged estimation process. Successively, the distribution moments of $\xi$ and the vector x are estimated in solving the inverse problem (see Chapter 5).

## 3. A General Optimization Approach

### 3.1. Optimal Solutions to Noisy Semi-Discretized Problems

#### 3.1.1. Identification Problems

Let us now establish an optimization approach for the approximate solution of identification and control problems, but in such a way that the solutions continuously depend upon the input data. In this Chapter 3, we only consider the strictly deterministic and non-Bayesian case (see Sec. 2.3.). The study of the present section deals with the semi-discretization model, i.e., the determination of Banach or Hilbert space elements from an m-dimensional noisy data vector z according to formulae (2.96) and (2.97). In the subsequent Sec. 3.2., we will turn to the full-discretization model, which is concerned with noisy discretized inverse problems in the sense of Definition 2.63. Furthermore, we have to distinguish the particularities in modelling identification and control problems. For problems arising in case of essentially coupled dimension numbers m and n (see for instance Sec. 2.2.4.), we refer to [209, Sec.3.3.]. Finally, Sec. 3.3. will complete this chapter with a study on the uncertainty of approximate solutions in the identification case.

Definition 3.1:

Consider the identification problem (2.1) with a discretization of the right-hand side $\mathcal{b}$ by a projector $Q^m$. Moreover, let the given data $z \in R^m$ be noisy according to the formulae (2.96) and (2.97). Then we call (2.1) with the quadruple $(m;Q^m;\delta;z)$ of intrinsic model ingredients semi-discretization model of the identification problem. Consequently, the problem of recovering $\varkappa \in \mathfrak{D}$ from $z \in R^m$ is termed a noisy semi-discretized identification problem. Furthermore, the closed set

$$\mathfrak{Z}_\delta := \{\varkappa \in \mathfrak{D} \subseteq B_1 : \|Q^m \mathcal{A}\varkappa - z\| \leq \delta\} \tag{3.1}$$

is said to be the set of elements which are compatible with the model (for short, set of compatible elements). o

Remark 3.2:

If $\mathcal{A}$ is linear, then the set of compatible elements $\mathfrak{Z}_\delta$ becomes convex. On the other hand, for a nonlinear operator $\mathcal{A}$, $\mathfrak{Z}_\delta$ may be a nonconvex and also nonconnected set. In view of the instability of problem (2.1), the set $\mathfrak{Z}_\delta$ may fail to be bounded even if $\delta$ is arbitrarily small. Therefore, a reasonable reconstruction of $\varkappa \in \mathfrak{D}$ from z has to use subjective a priori information in form of an evaluation functional in addition to the objective a priori information and data information, which together determine the set of compatible elements $\mathcal{Z}_\delta$. ○

<u>Assumption 3.3:</u>

In the sequel, let $B_0$ be a dense subset of $B_1$ and $\Pi: B_0 \to R$ a functional satisfying the following three conditions.

I. $\Pi(\varkappa) \geq 0$, for all $\varkappa \in B_0$, i.e., $\Pi$ is a nonnegative functional.

II. The level sets

$$W_c^{\Pi} := \{\varkappa \in B_0 : \Pi(\varkappa) \leq c\} \tag{3.2}$$

are, for any real number $c \geq 0$, empty or relatively compact subsets of $B_1$. For short, we say that $\Pi$ is a stabilizing functional.

III. There is an inner product $(.,.)_0$ on $B_0$ such that $B_0$ becomes a Hilbert space with a norm

$$\|\varkappa\|_0 = \sqrt{(\varkappa,\varkappa)_0} := \sqrt{\Pi(\varkappa)} \tag{3.3}$$

generated by the functional $\Pi$. In addition, assume that $\|.\|_0$ is a majorant norm with respect to $\|.\|_1$, i.e.

$$\|\varkappa\|_0^2 = \Pi(\varkappa) \geq \underline{c}^2 \cdot \|\varkappa\|_1^2 \text{ , for } \underline{c}>0 \text{ and all } \varkappa \in B_0. \tag{3.4}$$

○

<u>Lemma 3.4:</u>

Under the assumptions stated above the functional $\Pi: B_0 \to R$ is continuous on $B_0$ with respect to the norm $\|.\|_0$ and lower semi-continuous on $B_0$ with respect to the norm $\|.\|_1$. That is, for $\bar{\varkappa}, \varkappa^{(i)} \in B_0$, $i=1,2,\ldots$,

$$\lim_{i\to\infty} \Pi(\varkappa^{(i)}) = \Pi(\bar{\varkappa}) \qquad \text{as } \lim_{i\to\infty} \|\varkappa^{(i)} - \bar{\varkappa}\|_0 = 0 \tag{3.5}$$

and

$$\liminf_{i\to\infty} \Pi(\varkappa^{(i)}) \geq \Pi(\bar{\varkappa}) \qquad \text{as } \lim_{i\to\infty} \|\varkappa^{(i)} - \bar{\varkappa}\|_1 = 0 \text{ .} \tag{3.6}$$

The nonempty level sets $W_c^{\Pi}$ (see formula (3.2)) are compact (i.e., relatively compact and closed) in $B_1$. ○

<u>Proof:</u>

The condition (3.5) is a direct consequence of formula (3.3), since the norm functional is continuous on any Banach space (cf. [70, p.49]). In the sequel, we denote by the symbols $\xrightarrow[0]{}$, $\xrightarrow[1]{}$ the strong convergence of a sequence in the Banach spaces $B_0$ and $B_1$, respectively, and by $\xrightarrow[0]{\rightharpoonup}$, $\xrightarrow[1]{\rightharpoonup}$ the associated weak convergence (convergence of all bounded linear functionals) in $B_0$ and $B_1$. Firstly, we will show the closure of the level sets $W_c^{\Pi}$ in $B_1$. Let $\{\varkappa^{(i)}\}_{i=1}^{\infty} \subseteq W_c^{\Pi}$ with $\varkappa^{(i)} \xrightarrow[1]{} \bar{\varkappa} \in B_1$, then $\|\varkappa^{(i)}\|_0 \leq \sqrt{c}$, $i=1,2,\ldots$, and the sequence is weakly relatively compact in the Hilbert space $B_0$ (see [3, p.49]). Hence, there is a subsequence $\varkappa^{(i_j)} \xrightarrow[0]{} \varkappa' \in B_0$. Owing to formula (3.4) this implies $\varkappa^{(i_j)} \xrightarrow[1]{} \varkappa' \in B_0$. Namely, any bounded linear functional $\ell(\varkappa)$ on $B_1$ satisfies, for $\tilde{\varkappa} \in B_0$, the inequalities

$|\ell(\tilde{\varkappa})| \leq \|\ell\|_1 \cdot \|\tilde{\varkappa}\|_1 \leq \frac{\|\ell\|_1}{\underline{c}} \cdot \|\tilde{\varkappa}\|_0$ with $\|\ell\|_1 := \sup\limits_{0 \neq \varkappa \in B_1} \frac{|\ell(\varkappa)|}{\|\varkappa\|_1} < \infty$. Therefore,

$\|\ell\|_0 := \sup\limits_{0 \neq \tilde{\varkappa} \in B_0} \frac{|\ell(\tilde{\varkappa})|}{\|\tilde{\varkappa}\|_0} \leq \frac{\|\ell\|_1}{\underline{c}} < \infty$ and $\ell(\tilde{\varkappa})$ is also a bounded linear

functional on $B_0$. Consequently, $\varkappa' = \bar{\varkappa} \in B_0$, since $\varkappa^{(i)} \xrightarrow[1]{} \bar{\varkappa}$ implies $\varkappa^{(i)} \xrightarrow[1]{} \bar{\varkappa}$ and the weak limit of a weakly convergent sequence is uniquely determined. If we can prove the implication

$$\|\bar{\varkappa}\|_0 \leq \liminf_{j \to \infty} \|\varkappa^{(i_j)}\|_0 \qquad \text{as} \qquad \varkappa^{(i_j)} \xrightarrow[0]{} \bar{\varkappa}\ , \tag{3.7}$$

then $\|\bar{\varkappa}\|_0 \leq \sqrt{c}$ and $W_c^{\pi}$ is a closed and due to Assumption 3.3 also compact subset of $B_1$. Assume that $\liminf\limits_{j \to \infty} \|\varkappa^{(i_j)}\|_0^2 < \|\bar{\varkappa}\|_0$. From Riesz's theorem (see e.g. [3, p.36]) we derive the limiting condition $\lim\limits_{j \to \infty} (\varkappa^{(i_j)}, \varkappa^{(0)})_0 = (\bar{\varkappa}, \varkappa^{(0)})_0$, for all $\varkappa^{(0)} \in B_0$. Hence, we would obtain $\lim\limits_{j \to \infty} (\varkappa^{(i_j)} - \bar{\varkappa}, \bar{\varkappa})_0 = 0$ and thus, in view of the inequality

$0 \leq \|\varkappa^{(i_j)} - \bar{\varkappa}\|_0^2 = (\|\varkappa^{(i_j)}\|_0^2 - \|\bar{\varkappa}\|_0^2) - 2 \cdot (\varkappa^{(i_j)} - \bar{\varkappa}, \bar{\varkappa})_0$,

$\|\bar{\varkappa}\|_0^2 \leq \liminf\limits_{j \to \infty} \|\varkappa^{(i_j)}\|_0^2$. This is a contradiction. Therefore, $W_c^{\pi}$ is compact in $B_1$. Finally, let us prove the condition (3.6). Owing to formula (3.7) it suffices to show that, for $\bar{\varkappa}, \varkappa^{(i)} \in B_0$, $i = 1, 2, \ldots$, and $\varkappa^{(i)} \xrightarrow[1]{} \bar{\varkappa}$, there is a subsequence $\varkappa^{(i_j)} \xrightarrow[0]{} \bar{\varkappa}$. However, there exists a bounded subsequence $\{\varkappa^{(i_j)}\}_{j=1}^{\infty}$ with respect to the norm $\|.\|_0$ whenever $\liminf\limits_{i \to \infty} \|\varkappa^{(i)}\|_0 < \infty$. Then, according to the above ideas, we obtain $\varkappa^{(i_j)} \xrightarrow[0]{} \bar{\varkappa}$ as required. Note that in case of $\liminf\limits_{i \to \infty} \|\varkappa^{(i)}\|_0 = \infty$ formula (3.6) holds in a trivial manner. #

<u>Definition 3.5:</u>

Any global solution $\varkappa_{opt} \in B_0 \cap \mathcal{D}$ to the optimization problem

$$\underset{\varkappa \in B_0 \cap \mathcal{Z}_\delta}{\text{minimize}}\ \Pi(\varkappa) \tag{3.8}$$

is called optimal solution to the noisy semi-discretized identification problem (for short, sdi-optimal solution). The set of all sdi-optimal solutions will be denoted by $\mathcal{X}_{opt}$. ○

By using qualitative information about the solution to an inverse problem (2.1) stored in the set $\mathcal{D}$ and in the functional $\Pi$, we can exploit sdi-optimal solutions in order to complete our knowledge with respect to the quantitative behaviour of the solution element.

<u>Lemma 3.6:</u>

The optimization problem (3.8) is solvable whenever the set $B_0 \cap Z_\delta$ is nonempty. ○

<u>Proof:</u>

We consider the full inverse image $Z_\delta$ of the closed set $\{\tilde{y} \in R^m: \|\tilde{y}-z\| \leq \delta\}$ with respect to the composite operator $Q^m \mathcal{A}: \mathcal{D} \subseteq B_1 \to R^m$. Since $Q^m \mathcal{A}$ is continuous and $\mathcal{D}$ is closed in $B_1$, $Z_\delta$ becomes closed in $B_1$. Therefore, the coming lemma yields the assertion of our lemma under consideration. #

<u>Lemma 3.7:</u>

Let $\mathcal{U}$ be a closed subset of $B_1$ and $\Pi$ a functional satisfying the Assumption 3.3. Then there exists an element $\varkappa' \in B_0 \cap \mathcal{U}$ with $\Pi(\varkappa') = \inf_{\varkappa \in B_0 \cap \mathcal{U}} \Pi(\varkappa)$, i.e., there is a minimiser of $\Pi$ in the set $B_0 \cap \mathcal{U}$ whenever $B_0 \cap \mathcal{U}$ is nonempty. ○

<u>Proof:</u>

As we know from Lemma 3.4, the level set $W^{\Pi}_{\Pi(\varkappa^{(0)})} = \{\varkappa \in B_0: \Pi(\varkappa) \leq \Pi(\varkappa^{(0)})\}$ is compact in $B_1$ for an arbitrarily chosen element $\varkappa^{(0)} \in B_0 \cap \mathcal{U}$. Moreover, in view of the closure of $\mathcal{U}$ in $B_1$, the intersection $W^{\Pi}_{\Pi(\varkappa^{(0)})} \cap \mathcal{U}$ is also compact in $B_1$. We have

$\inf_{\varkappa \in B_0 \cap \mathcal{U}} \Pi(\varkappa) = \inf_{\varkappa \in W^{\Pi}_{\Pi(\varkappa^{(0)})} \cap \mathcal{U}} \Pi(\varkappa)$ with $\Pi$ lower semi-continuous on $B_0$ according to $\|\cdot\|_1$. For any minimizing sequence $\{\varkappa^{(i)}\}_{i=1}^{\infty} \subseteq W^{\Pi}_{\Pi(\varkappa^{(0)})} \cap \mathcal{U}$ satisfying $\lim_{i \to \infty} \Pi(\varkappa^{(i)}) = \inf_{\varkappa \in B_0 \cap \mathcal{U}} \Pi(\varkappa)$, there is an element $\varkappa' \in W^{\Pi}_{\Pi(\varkappa^{(0)})} \cap \mathcal{U}$ with $\varkappa^{(i_j)} \xrightarrow[1]{} \varkappa'$. From the lower semi-continuity of $\Pi$ it follows $\Pi(\varkappa') \leq \inf_{\varkappa \in B_0 \cap \mathcal{U}} \Pi(\varkappa)$ and thus $\Pi(\varkappa') = \min_{\varkappa \in B_0 \cap \mathcal{U}} \Pi(\varkappa)$. #

Remark 3.8:

The above introduction of a stabilizing functional and of associated sdi-optimal solutions follows the concept of TIKHONOV and ARSENIN [434, Chap.2]. Provided that the requirements I., II. and III. of Assumption 3.3 are satisfied, Lemma 3.7 is already proven in [434, p. 62] without mentioning the lower semi-continuity of $\Pi$. This proof essentially uses the Hilbert space properties of $B_0$ in order to show that any minimizing sequence is also a Cauchy sequence. Then the minimizer of $\Pi$ subject to $B_0 \cap \mathcal{U}$ is attained at least at one point, since $B_0$ is a complete space. ○

Remark 3.9:

Stabilizing functionals have also been presented by VASILEV [460, Chap.2]. VASILEV's requirements on $\Pi$ are weaker compared to those of Assumption 3.3, but they allow similar conclusions. Instead of condition III., the lower semi-continuity of $\Pi$ on $B_0$ with respect to $\|.\|_1$ is immediately required. However, condition II. must be formulated in a stronger form as follows:

II". The level sets $W_c^{\Pi}$ according to (3.2) are, for any real number $c \geq 0$, empty or compact subsets of $B_1$.

It may be easier to show that condition III. holds than to prove the lower semi-continuity of $\Pi$. ○

Examples of stabilizing functionals are given in [460, p.166]. We only mention three frequently used cases.

Remark 3.10:

a. The nonnegative functional (2.95) satisfies the requirements of Assumption 3.3 for $\tilde{p}_0 > 0$, $B_0 := B_1 := R^n$. Condition II. is evident from the fact that any bounded subset of $R^n$ gets relatively compact. Provided $(x^{(1)}, x^{(2)})_0 := (Mx^{(1)}, x^{(2)})$, $\|x\|_1 := \|x\|$, $M := \tilde{p}_0 M_0 + \tilde{p}_1 M_1$, $M_0 := I \in R^{n \times n}$ (unity matrix), $M_2 := \begin{bmatrix} 1 & -1 & & & & \\ -1 & 2 & -1 & & 0 & \\ & -1 & 2 & -1 & & \\ & & \ddots & \ddots & \ddots & \\ & 0 & & -1 & 2 & -1 \\ & & & & -1 & 1 \end{bmatrix} \in R^{n \times n}$ (symmetric positive semi-definite tridiagonal matrix), we have $\|x\|_0^2 = \Omega(x) \geq \tilde{p}_0 \cdot \|x\|_1^2$ for all $x \in R^n$ whenever $\tilde{p}_0 > 0$. Hence, condition III. is valid. In the sequel, stabilizing functionals on $R^n$ will always be denoted by $\Omega$ in order to characterize a finite dimensional domain, whereas the symbol $\Pi$ is used when the domain $B_0$ is infinite dimensional.

b. The nonnegative functional (2.94) satisfies the Assumption 3.3 for positive values $p_0$ and $p_1$, $B_1 := C[0,1]$, $B_0 := W_2^{(1)}[0,1] \subseteq B_1$ with a norm $\|.\|_0$ equivalent to (2.31) and generated by

$$(\varkappa^{(1)},\varkappa^{(2)})_0:=p_0\cdot\int_0^1 x^{(1)}(t)\cdot x^{(2)}(t)dt+ p_1\cdot\int_0^1 (x^{(1)})'(t)\cdot(x^{(2)})'(t)dt.$$

Here, we understand by $W_2^{(1)}[0,1]$ the set of continuous representatives of functions possessing a square integrable generalized derivative (see [434, 2nd Russian ed., p. 130]). The compactness of $W_c^{\pi}$ in $B_1$ is a consequence of the compact embedding of $W_2^{(1)}[0,1]$ into $C[0,1]$ (cf. [402, Vol.5, p. 317]). If the norm $\|.\|_1$ is weakened to $B_1:=L_q[0,1]$, $1\leq q<\infty$ with

$$\|\varkappa\|_{L_q[0,1]}=\left(\int_0^1 |x(t)|^q\cdot dt\right)^{\frac{1}{q}}, \tag{3.9}$$

then the stabilizing property of $\pi$ is maintained.

c. Let $B_1:=B_2:=H_1$ be a Hilbert space with inner product $(.,.)_1$. Moreover, let $\mathcal{B}:\mathcal{D}(\mathcal{B})\subseteq H_1\to H_1$ ($\mathcal{D}(\mathcal{B})$- domain of $\mathcal{B}$) be a linear densely defined selfadjoint positive definite operator with eigenvalues $0<\lambda_1\leq\lambda_2\leq\ldots\leq\lambda_i\leq\lambda_{i+1}\leq\ldots\to\infty$ as $i\to\infty$ and an associated complete orthonormal eigensystem $\{u^{(i)}\}_{i=1}^{\infty}\subseteq H_1$. The linear inverse operator $\mathcal{B}^{-1}: H_1\to H_1$ of $\mathcal{B}$ is compact. Then, the nonnegative functional

$$\pi(\varkappa):=(\mathcal{B}\varkappa,\varkappa)_1=\sum_{i=1}^{\infty}\lambda_i(\varkappa,u^{(i)})_1^2\geq\lambda_1\cdot\sum_{i=1}^{\infty}(\varkappa,u^{(i)})_1^2=\lambda_1\|\varkappa\|_1^2 \tag{3.10}$$

satisfies the Assumption 3.3. Here, the inner product $(\varkappa^{(1)},\varkappa^{(2)})_0:=(\mathcal{B}^{1/2}\varkappa,\mathcal{B}^{1/2}\varkappa)_1$ takes on finite values on the domain

$$\mathcal{D}(\mathcal{B}^{1/2})=\{\varkappa=\sum_{i=1}^{\infty}(\varkappa,u^{(i)})_1 u^{(i)}\in B_1:\ \sum_{i=1}^{\infty}\lambda_i(\varkappa,u^{(i)})_1^2<\infty\}:=B_0:=H_0$$

of the square root $\mathcal{B}^{1/2}$. The Hilbert space $B_0:=H_0$ is the energetic space (see [302, p. 773]) of $H_1$ with respect to the unbounded linear operator $\mathcal{B}$. Due to

$W_c^{\pi}=\mathcal{B}^{-1/2}\cdot V_c$ with $V_c:=\{b\in H_1:\|b\|_1^2\leq c\}$ bounded in $H_1$ and $\mathcal{B}^{-1/2}: H_1\to H_1$ compact, it is evident that $W_c^{\pi}$ is relatively compact in $H_1$. ○

We are now going to study some properties of sdi-optimal solutions. Specifically, the stability of such solutions with respect to perturbations in the input data and the asymptotic behaviour as $m\to\infty$ are of importance. Firstly, we generalize the well-posedness definition (see Definition 2.6) to the case of a general constrained optimization problem.

**Definition 3.11:**

An optimization problem on the Banach space $B_1$

$$\underset{\varkappa \in B_0 \cap \mathcal{U}}{\text{minimize}} \; f(\varkappa) \tag{3.11}$$

with an objective functional $f : B_0 \subseteq B_1 \to R$ and a nonempty subset $\mathcal{U}$ of $B_1$ is called well-posed if the following conditions are both satisfied:

IV. $B_0 \cap \mathcal{U} \neq \emptyset$, $\inf_{\varkappa \in B_0 \cap \mathcal{U}} f(\varkappa) > -\infty$ and $\mathcal{X}_{\mathcal{U}} := \{\varkappa^{(0)} \in B_0 \cap \mathcal{U} : f(\varkappa^{(0)}) = \inf_{\varkappa \in B_0 \cap \mathcal{U}} f(\varkappa)\} \neq \emptyset$ (existence of optimal solutions).

V. Given $\varepsilon > 0$, we can choose $\delta = \delta(\varepsilon) > 0$ so that $\text{dist}(\varkappa, \mathcal{X}_{\mathcal{U}}) := \inf_{\varkappa^{(1)} \in \mathcal{X}_{\mathcal{U}}} \|\varkappa - \varkappa^{(1)}\|_1 < \varepsilon$ whenever $\varkappa \in B_0 \cap \mathcal{U}$ and $f(\varkappa) < \inf_{\varkappa^{(2)} \in B_0 \cap \mathcal{U}} f(\varkappa^{(2)}) + \delta$

(stability of optimal solutions with respect to the objective functional).

Otherwise the problem (3.11) is called ill-posed. ○

**Theorem 3.12:**

The optimization problem (3.8) aimed at finding sdi-optimal solutions is well-posed on the Banach space $B_1$ whenever $B_0 \cap \mathcal{Z}_\delta$ is a nonempty set. ○

**Proof:**

In view of Lemma 3.6, condition IV. is satisfied. Provided that there was a sequence $\{\varkappa^{(i)}\}_{i=1}^{\infty} \subseteq B_0 \cap \mathcal{Z}_\delta$ with $\lim_{i \to \infty} \pi(\varkappa^{(i)}) = \pi(\varkappa_{opt})$ and $\text{dist}(\varkappa^{(i)}, \mathcal{X}_{opt}) \geq \varepsilon$, $i = 1, 2, \ldots$, we would obtain a contradiction. Namely, owing to condition II. of Assumption 3.3 and Lemma 3.4, $\{\varkappa^{(i)}\}_{i=1}^{\infty}$ is a relatively compact subset of $B_1$ and any point of accumulation $\varkappa'$ belongs to the set $B_0 \cap \mathcal{Z}_\delta$. Hence, the lower semi-continuity of $\pi$ provides $\pi(\varkappa') \leq \pi(\varkappa_{opt})$ and $\varkappa' \in \mathcal{X}_{opt}$. Therefore, there is a subsequence $\{\varkappa^{(i_j)}\}_{j=1}^{\infty}$ such that $\lim_{j \to \infty} \text{dist}(\varkappa^{(i_j)}, \mathcal{X}_{opt}) = 0$, violating the above assumptions. #

**Definition 3.13:**

The set $\mathcal{U} \in 2^{B_1}$ of all points of accumulation of the sequence $\{\mathcal{U}^{(i)}\}_{i=1}^{\infty} \subseteq 2^{B_1}$ with respect to the norm $\|.\|_1$, i.e.,

$$\mathcal{U} := \limsup_{i \to \infty} \mathcal{U}^{(i)} := \{\varkappa \in B_1 : \varkappa^{(i_j)} \xrightarrow[1]{} \varkappa \text{ for a subsequence } \varkappa^{(i_j)} \in \mathcal{U}^{(i_j)}\}, \tag{3.12}$$

is the superior limit of the subset sequence (see e.g. [425]; for limits of set sequences in general, see also [324]). ○

Definition 3.14:

Given an objective functional $f$, the optimisation problem (3.11) on $B_1$ is called stable at $\mathcal{U}\in\mathcal{C}$ with respect to a family $\mathcal{C}\subseteq 2^{B_1}$ of constraints if the following conditions are both satisfied:

VI. The problem (3.11) is well-posed in the sense of Definition 3.11 for any $\tilde{\mathcal{U}}\in\mathcal{C}$.

VII. We have $\emptyset \neq \limsup_{i\to\infty} \mathcal{X}_{\mathcal{U}^{(i)}} \subseteq \mathcal{X}_{\mathcal{U}}$

whenever $\mathcal{U}^{(i)}\in\mathcal{C}$ and $\limsup_{i\to\infty} \mathcal{U}^{(i)} \subseteq \mathcal{U}$ .

If the above conditions are satisfied, we also say that the optimal solutions to (3.11) at $\mathcal{U}$ stably depend on the constraints with respect to the family $\mathcal{C}$. o

Theorem 3.15:

The optimisation problem on $B_1$

$$\underset{\varkappa\in B_0\cap\mathcal{U}}{\text{minimise}}\ \Pi(\varkappa) \tag{3.13}$$

is stable at $\mathcal{U}$ with respect to any family $\mathcal{C}$ containing closed subsets $\tilde{\mathcal{U}}$ of $B_1$ whenever the intersection of this family and $\mathcal{X}_{\mathcal{U}}$ is nonempty, i.e.,

$$\mathcal{X}_{\mathcal{U}}\cap\bigcap_{\tilde{\mathcal{U}}\in\mathcal{C}}\tilde{\mathcal{U}}\neq\emptyset .\ \text{o} \tag{3.14}$$

Proof:

The condition VI. is satisfied due to Theorem 3.12 if $Z_\delta$ is replaced by an arbitrarily chosen nonempty closed subset $\tilde{\mathcal{U}}$ of $B_1$. Now, let $\limsup_{i\to\infty} \mathcal{U}^{(i)}\subseteq\mathcal{U}$ and $\tilde{\varkappa}\in\mathcal{X}_{\mathcal{U}}\cap\bigcap_{i=1}^{\infty}\mathcal{U}^{(i)}$ , where $\mathcal{U}^{(i)}\subseteq B_1$ are closed subsets for all i. We denote by $\varkappa^{(i)}$, i=1,2,..., solutions to the problems

$$\underset{\varkappa\in B_0\cap\mathcal{U}^{(i)}}{\text{minimise}}\ \Pi(\varkappa)\ ,\ i=1,2,..., \tag{3.15}$$

respectively. Then, $\Pi(\varkappa^{(i)})\leq\Pi(\tilde{\varkappa})$, for all i, and owing to the compactness of $W^{\Pi}_{\Pi(\tilde{\varkappa})}$ (see Lemma 3.4) in $B_1$, we obtain $\limsup_{i\to\infty}\mathcal{X}_{\mathcal{U}^{(i)}}\neq\emptyset$, i.e., there is a point of accumulation $\varkappa'\in B_1$ of this sequence $\{\varkappa^{(i)}\}_{i=1}^{\infty}$. Moreover, $\varkappa'\in B_0$ and from $\limsup_{i\to\infty}\mathcal{U}^{(i)}\subseteq\mathcal{U}$ it follows $\varkappa'\in B_0\cap\mathcal{U}$ . By formula (3.6) we finally obtain $\Pi(\varkappa')\leq\liminf_{i\to\infty}\Pi(\varkappa^{(i)})\leq\Pi(\tilde{\varkappa})$ and therefore $\varkappa'\in\mathcal{X}_{\mathcal{U}}$ as well as $\limsup_{i\to\infty}\mathcal{X}_{\mathcal{U}^{(i)}}\subseteq\mathcal{X}_{\mathcal{U}}$ . #

Theorem 3.15 is useful for studying the stability of sdi-optimal solutions with respect to perturbations in z and $\delta$. Assume that the data vector z is superposed by a small additional error which arises if the measured numbers are prepared for the computer. Let $(m\,;Q^m\,;\delta_i\,;z^{(i)})$, i=1,2,..., be a sequence of model quadruples (see Definition 3.1) with

$$\|z^{(i)}-z\|\le \sigma_i\,,\quad \delta_i:=\delta+\sigma_i,\ i=1,2,\dots, \tag{3.16}$$

such that $\mathcal{Z}^{(i)}\supseteq\mathcal{Z}_\delta$, for all i, and

$$\mathcal{Z}^{(i)}:=\{\varkappa\in\mathcal{D}\subseteq B_1:\ \|Q^m\mathcal{A}\varkappa-z^{(i)}\|\le\delta_i\}\ . \tag{3.17}$$

This implies $\mathcal{X}_{opt}=\mathcal{X}_{opt}\cap\bigcap_{i=1}^{\infty}\mathcal{Z}^{(i)}$, and optimal solutions to (3.8) are stable at $\mathcal{Z}_\delta$ with respect to constraints $\mathcal{Z}^{(i)}$ of the form (3.16) and (3.17).

<u>Theorem 3.16:</u>
Provided that (3.16) holds with $\lim\limits_{i\to\infty}\sigma_i=0$, we obtain $\limsup\limits_{i\to\infty}\mathcal{Z}^{(i)}\subseteq\mathcal{Z}_\delta$, and the sets $\mathcal{X}^{(i)}$, i=1,2,..., of solutions to

$$\underset{\varkappa\in B_0\cap\mathcal{Z}^{(i)}}{\text{minimize}}\ \Pi(\varkappa),\ i=1,2,\dots, \tag{3.18}$$

satisfy the condition

$$\emptyset\ne\limsup_{i\to\infty}\mathcal{X}^{(i)}\subseteq\mathcal{X}_{opt}\ .\ \circ \tag{3.19}$$

<u>Proof:</u>
This theorem is a corollary of Theorem 3.15 if one can show that $\limsup\limits_{i\to\infty}\mathcal{Z}^{(i)}\subseteq\mathcal{Z}_\delta$ as $\lim\limits_{i\to\infty}\sigma_i=0$. However, this inclusion is evident from the inequalities $\|Q^m\mathcal{A}\varkappa^{(i_j)}-z\|\le\|Q^m\mathcal{A}\varkappa^{(i_j)}-z^{(i_j)}\|+\|z^{(i_j)}-z\|\le \delta_{i_j}+\sigma_{i_j}=\delta+2\cdot\sigma_{i_j}$. Namely, for $\varkappa^{(i_j)}\in\mathcal{Z}^{(i_j)}$ and $\varkappa^{(i_j)}\xrightarrow[1]{}\varkappa'$, we obtain $\|Q^m\mathcal{A}\varkappa'-z\|\le\delta$ with $\lim\limits_{j\to\infty}\sigma_{i_j}=0$ in view of the continuity of the operator $Q^m\mathcal{A}:\mathcal{D}\subseteq B_1\longrightarrow R^m$. #

As the above theorem shows, optimal solutions to problems with perturbed input data are converging to sdi-optimal solutions $\varkappa_{opt}$ if the additional perturbations tend to zero.

Now we are going to study the relations between the set of exact solutions

$$\mathcal{Z}^*:=\{\varkappa\in\mathcal{D}\subseteq B_1:\ \mathcal{A}\varkappa=\ell\} \tag{3.20}$$

and sdi-optimal solutions if the number of data m tends to infinity and the measurement errors become small. Here, let $(m_i;Q^{m_i};\delta_i;z^{(i)})$, i=1,2,..., be a sequence of model quadruples with $m_1<m_2<\dots<m_i<m_{i+1}<\dots\to\infty$, $\lim\limits_{i\to\infty}\delta_i=0$ and

$$\|Q^{m_i} b - z^{(i)}\| \leq \delta_i \ , \quad i=1,2,\ldots \ . \tag{3.21}$$

**Theorem 3.17:**

Let $B_0 \cap \mathcal{Z}^* \neq \emptyset$ . Moreover, let $\|.\|_*$ be a norm in $R^{m_i}$ such that

$$\lim_{i\to\infty} \|Q^{m_i}\tilde{b}\|_* = \|\tilde{b}\|_2 \ , \text{ for all } \tilde{b} \in B_2 \ , \tag{3.22}$$

and this convergence is uniform on any compact $\mathcal{K} \subseteq B_2$, i.e., for all $\varepsilon > 0$ there is an $m_0 = m_0(\varepsilon, \mathcal{K})$ with $\left| \|Q^{m_i}\tilde{b}\|_* - \|\tilde{b}\|_2 \right| < \varepsilon$ whenever $m_i > m_0$ and $\tilde{b} \in \mathcal{K}$. Furthermore, assume

$$\|v\|_* \leq \varphi(m_i) \cdot \|v\| \ , \text{ for } \varphi(m_i) > 0 \text{ and all } v \in R^{m_i}. \tag{3.23}$$

If

$$\lim_{i\to\infty} \delta_i \cdot \varphi(m_i) = 0 \ , \tag{3.24}$$

$$\mathcal{Z}^{(i)} := \{ \varkappa \in \mathcal{D} \subseteq B_1 : \|Q^{m_i} \mathcal{A}\varkappa - z^{(i)}\| \leq \delta_i \} \tag{3.25}$$

and we denote by $\mathcal{X}^*$ and $\mathcal{X}^{(i)}$, $i=1,2,\ldots$, the sets of solutions to

$$\underset{\varkappa \in B_0 \cap \mathcal{Z}^*}{\text{minimize}} \ \Pi(\varkappa) \tag{3.26}$$

and

$$\underset{\varkappa \in B_0 \cap \mathcal{Z}^{(i)}}{\text{minimize}} \ \widetilde{\Pi}(\varkappa) \ , \quad i=1,2,\ldots \ , \tag{3.27}$$

respectively, then we have $\limsup_{i\to\infty} \mathcal{Z}^{(i)} \subseteq \mathcal{Z}^*$ and

$$\emptyset \neq \limsup_{i\to\infty} \mathcal{X}^{(i)} \subseteq \mathcal{X}^* \ . \quad \circ \tag{3.28}$$

**Proof:**

Obviously, in view of formula (3.21), we have for all $i$ the inclusion $\mathcal{Z}^* \subseteq \mathcal{Z}^{(i)}$. Thus, $\mathcal{X}^* \cap \bigcap_{i=1}^{\infty} \mathcal{Z}^{(i)} \neq \emptyset$ and the optimal solutions to (3.26) are stable at $\mathcal{Z}^*$ with respect to constraint perturbations of the form $\mathcal{Z}^{(i)}$ (see Definition 3.14). Then the formula (3.28) is a consequence of Theorem 3.15 whenever $\limsup_{i\to\infty} \mathcal{Z}^{(i)} \subseteq \mathcal{Z}^*$ can be proven. Let be $\varkappa^{(i_j)} \in \mathcal{Z}^{(i_j)}$ with $\varkappa^{(i_j)} \xrightarrow[1]{} \varkappa'$. We only have to show $\varkappa' \in \mathcal{Z}^*$. Now,

$$\|Q^{m_{i_j}} \mathcal{A}\varkappa^{(i_j)} - Q^{m_{i_j}} b\|_* \leq \varphi(m_{i_j}) \cdot \|Q^{m_{i_j}} \mathcal{A}\varkappa^{(i_j)} - Q^{m_{i_j}} b\| \leq$$

$$\leq \varphi(m_{i_j}) ( \|Q^{m_{i_j}} \mathcal{A}\varkappa^{(i_j)} - z^{(i_j)}\| + \|Q^{m_{i_j}} b - z^{(i_j)}\| ) \leq 2 \cdot \varphi(m_{i_j}) \cdot \delta_{i_j}$$ and

thus $\lim_{j\to\infty} \|Q^{m_{i_j}}(\mathcal{A}\varkappa^{(i_j)} - b)\|_* = 0$ . We will derive that this implies $\lim_{j\to\infty} \|\mathcal{A}\varkappa^{(i_j)} - b\|_2 = 0$ . Then, $\mathcal{A}\varkappa' = b$ and $\varkappa' \in \mathcal{Z}^*$ in view of the closure of $\mathcal{D}$ in $B_1$. Assume the sequence $\tilde{b}^{(j)} := \mathcal{A}\varkappa^{(i_j)} - b$ to be converging to $b' := \mathcal{A}x' - b$ with respect to $\|.\|_2$. Moreover, let $\varepsilon > 0$ be a fixed value. The union set $\mathcal{K} := \{\tilde{b}^{(j)}\}_{j=1}^{\infty} \cup b'$ is a compact subset of $B_2$ and we have integers $j_0$ and $j_1$ such that

$$\left| \|Q^{m_{i_j}} \tilde{b}\|_* - \|\tilde{b}\|_2 \right| < \frac{\varepsilon}{2} \text{ as } j \geq j_0,\ \tilde{b} \in \mathcal{K} \text{ and } \|Q^{m_{i_j}} \tilde{b}^{(j)}\|_* < \frac{\varepsilon}{2} \text{ as } j \geq j_1.$$

Finally, we obtain, for $j \geq \max(j_0, j_1)$, $\|\mathcal{A}x^{(i_j)} - b\|_2 = \|\tilde{b}^{(j)}\|_2 < \frac{\varepsilon}{2} + \|Q^{m_{i_j}} \tilde{b}^{(j)}\|_* \leq \frac{\varepsilon}{2} + \frac{\varepsilon}{2} = \varepsilon$ and thus $x' \in Z^*$. #

**Remark 3.18:**
Provided that (3.22) holds, a sufficient condition for the uniformity of this convergence on any compact set $\mathcal{K} \subseteq B_2$ is the monotonicity requirement

$$\left| \|Q^{m_i} \tilde{b}\|_* - \|\tilde{b}\|_2 \right| \geq \left| \|Q^{m_j} \tilde{b}\|_* - \|\tilde{b}\|_2 \right|, \tag{3.29}$$

for sufficiently large $m_0$ and all $m_0 \leq m_i < m_j$, $\tilde{b} \in B_2$. This remark is a consequence of assertions due to the concept of quasiuniform convergence (see ALEXANDROV [5, 2nd. Appendix], also [209, p.24]) as a generalization of Dini's theorem (see e.g. [129, Vol.2]). $\circ$

**Remark 3.19:**
a. If $Q^m$ is chosen according to formula (2.33) on the Hilbert space $B_2 := H_2$ and $\|b\|_* := \|b\|$, $\varphi(m) \equiv 1$, then due to Parseval's identity and for a complete orthonormal system $\{f^{(j)}\}_{j=1}^{\infty}$ we have

$$\|b\|_2^2 = \sum_{j=1}^{\infty} (b, f^{(j)})_2^2 = \lim_{m \to \infty} \|Q^m b\|_*^2 . \tag{3.30}$$

Hence, it follows (3.22) and via (3.29) the uniformity of this convergence on any compact $\mathcal{K} \subseteq B_2$ becomes clear. In this Hilbert space case, (3.24) has the form $\lim_{i \to \infty} \delta_i = 0$. If the measurement error norm bound tends to zero as $m \to \infty$, then exact solutions of the problem (2.1) are approximated by sdi-optimal solutions arbitrarily well. However, if we only know that any measurement error component vanishes as $m \to \infty$, a convergence of subsequences in the sense of formula (3.28) must not be expected.

b. Let $B_2 := C[0,1]$, $Q^m$ chosen according to (2.51), i.e., $Q^m b := (b(\frac{1}{m}), b(\frac{2}{m}), \ldots, b(1))^T$. Moreover, let be $\|b\|_* := \max_{1 \leq j \leq m} |b_j|$, then $\|b\|_* \leq \|b\|$, $\varphi(m) \equiv 1$ and $\left| \|Q^m b\|_* - \|b\|_2 \right| \leq \max_{\substack{s, s' \in [0,1] \\ |s - s'| \leq \frac{1}{m}}} |b(s) - b(s')|$. We introduce the modulus of continuity for a continuous function $b \in C[0,1]$ as follows:

$$\omega_{b}^{C[0,1]}(\varrho) := \sup_{\substack{s,s'\in[0,1] \\ |s-s'|\leq\varrho}} |b(s)-b(s')| . \tag{3.31}$$

It is well-known that $\omega_{b}^{C[0,1]}(\varrho)$ is a nonnegative and monotonically nondecreasing function for $\varrho\geq 0$. The modulus of continuity can be proven to be continuous with respect to $b\in C[0,1]$, for fixed $\varrho$, as well as with respect to $\varrho$, for given $b$. Moreover, $\lim_{\varrho\to 0} \omega_{b}^{C[0,1]}(\varrho)=0$ and in view of Arzela's theorem (see [39,p.164]) $\omega_{b}^{C[0,1]}(\varrho)\leq\omega_{\mathcal{K}}^{C[0,1]}(\varrho)$, i.e., for any compact set $\mathcal{K}\subseteq C[0,1]$ there is a uniform upper bound function $\omega_{\mathcal{K}}^{C[0,1]}(\varrho)$. This implies (3.22) in a uniform manner on any compact set $\mathcal{K}\subseteq C[0,1]$.

c. Assume that $B_2 := C^{(1)}[0,1]$ is the set of continuously differentiable functions (in the classical sense) on the interval $[0,1]$ and

$$\|b\|_{C^{(1)}[0,1]} := \max_{s\in[0,1]} |b(s)| + \max_{s\in[0,1]} |b'(s)| \tag{3.32}$$

is the norm of this Banach space. Let $Q^m$ be defined as considered in paragraph b. of this remark. Furthermore, let

$$\|b\|_{*} := \max_{1\leq j\leq m} |b_j| + \max_{\substack{j''<j' \\ 2\leq j'\leq m,\ 1\leq j''\leq m-1}} \frac{|b_{j'}-b_{j''}|}{|(j'-j'')\cdot\frac{1}{m}|} \leq \|b\| + m\cdot\max_{j',j''} (|b_{j'}|+|b_{j''}|) \leq$$

$\leq (2m+1)\cdot\|b\|$, i.e., $\varphi(m)=2m+1$. Then we have $\big|\|Q^m b\|_{*}-\|b\|_2\big| =$

$$\Big|\max_{1\leq j\leq m} |b(\tfrac{j}{m})| - \max_{s\in[0,1]} |b(s)| + \max_{j''<j'} |b'(\eta_{j'}^{j''})| - \max_{s\in[0,1]} |b'(s)|\Big| \leq$$

$$\omega_{b}^{C[0,1]}(\tfrac{1}{m}) + \omega_{b'}^{C[0,1]}(\tfrac{1}{m}) , \text{ where } \tfrac{j''}{m}<\eta_{j'}^{j''}<\tfrac{j'}{m},\ b'=\langle b'(s),\ 0\leq s\leq 1\rangle .$$

The ideas of paragraph b. again supply (3.22) with the required uniform convergence on compact subsets of $C^{(1)}[0,1]$. Note that formula (3.24) corresponds to $\lim_{i\to\infty} m_i\cdot\delta_i = 0$. In order to get convergence of sdi-optimal solutions to exact solutions of problem (2.1), $\delta_i$ has to fall faster, for $m_i\to\infty$, than in case of $C[0,1]$. This is a natural requirement, since $\|.\|_{C^{(1)}[0,1]}$ is a majorant norm with respect to $\|.\|_{C[0,1]}$. ○

### 3.1.2. Control Problems

**Definition 3.20:**

Consider the control problem (2.3) with a discretization of the desired element $b$ by a projector $Q^m$. Let the given data $z\in R^m$ be noisy subject to (2.96) and (2.97). Then, (2.3) with the quadruple $(m;Q^m;\delta;z)$ of intrinsic model ingredients is called semi-discretization model of the control problem. Consequently, the problem of finding an approximate control element $\varkappa\in\mathfrak{D}$ from $z\in R^m$ is termed a semi-discretized control problem. ○

Again, we have to consider a stabilizing functional $\Pi$ following the Assumption 3.3 with level sets $W_c^{\Pi}:=\{\varkappa\in B_0 : \Pi(\varkappa)\leq c\}$ compact in $B_1$.

Definition 3.21:

For given $c>0$, any global solution $\varkappa_c\in W_c^{\Pi}\cap\mathfrak{D}$ to the optimization problem

$$\underset{\varkappa\in W_c^{\Pi}\cap\mathfrak{D}}{\text{minimize}} \; \|Q^m\mathcal{A}\varkappa - z\|^2 \tag{3.33}$$

is called optimal solution to the noisy semi-discretized control problem (for short, sdc-optimal solution). The set of all sdc-optimal solutions to (3.33) will be denoted by $\mathcal{X}_c$. ○

Lemma 3.22:

The optimization problem (3.33) is solvable whenever the set $W_c^{\Pi}\cap\mathfrak{D}$ is nonempty. ○

Proof:

The Euclidean norm square $\|Q^m\mathcal{A}\varkappa-z\|^2$ depends upon $\varkappa\in\mathfrak{D}$ continuously with respect to $\|.\|_1$. Then the Weierstrass theorem (see e.g. [284, p.154]) applies to the optimization problem (3.33), since $W_c^{\Pi}\cap\mathfrak{D}$ is compact (relatively compact and closed) in $B_1$. Therefore, $\|Q^m\mathcal{A}\varkappa-z\|^2$ has at least one minimizer $\varkappa_c$ in $W_c^{\Pi}\cap\mathfrak{D}$. #

Theorem 3.23:

The optimization problem (3.33) of finding sdc-optimal solutions is well-posed on the Banach space $B_1$ whenever $W_c^{\Pi}\cap\mathfrak{D}$ is a nonempty set. ○

Proof:

The conditions IV. and V. of Definition 3.11 will be proven to be satisfied. Note that $f(\varkappa)$, $B_0$, $B_1$ and $\mathcal{U}$ in formulating (3.11) correspond to $\|Q^m\mathcal{A}\varkappa-z\|^2$, $B_1$, $B_1$ and $W_c^{\Pi}\cap\mathfrak{D}$, respectively, in (3.33). Owing to Lemma 3.22, condition IV. is satisfied. Provided that there was a sequence $\{\varkappa^{(i)}\}_{i=1}^{\infty}\subseteq W_c^{\Pi}\cap\mathfrak{D}$ with $\lim_{i\to\infty}\|Q^m\mathcal{A}\varkappa^{(i)}-z\|^2=\|Q^m\mathcal{A}\varkappa_c-z\|^2$ and $\text{dist}(\varkappa^{(i)},\mathcal{X}_c)\geq\varepsilon$, for all i, we would obtain a contradiction. Obviously, for compactness reasons, there is a subsequence $\varkappa^{(i_j)}\xrightarrow[1]{}\varkappa'\in W_c^{\Pi}\cap\mathfrak{D}$ of $\{\varkappa^{(i)}\}_{i=1}^{\infty}$. With respect to the continuity of the objective functional, it follows $\|Q^m\mathcal{A}\varkappa'-z\|^2=\|Q^m\mathcal{A}\varkappa_c-z\|^2$ and thus $\varkappa'\in\mathcal{X}_c$, violating the above assumption. #

From Theorem 3.23 we have learned that the set $\mathcal{X}_c$ continuously depends on z, i.e., small changes of z will cause small changes in $\mathcal{X}_c$. If we ask for the stability of $\mathcal{X}_c$ with respect to c, this question is concerned with the stability of the optimization problem (3.33) with respect to constraints in the sense of Definition 3.14. Analogously to Theorem 3.16 and based on Theorem 3.15 we can derive that $\limsup_{i\to\infty} \mathcal{X}_{c_i} \subseteq \mathcal{X}_c$ as $\lim_{i\to\infty} c_i = c$ whenever $c < c_i$, $i=1,2,\ldots$, and $W_c^\pi \cap \mathcal{D} \neq 0$. That is, sdc-optimal solutions stably depend on the constraints at $c > 0$ with respect to a family $\mathcal{C}$ (see Definition 3.14) of the form $W_{\tilde{c}}^\pi \cap \mathcal{D}$ including $W_c^\pi \cap \mathcal{D}$.

At the end of this Section 3.1., we have to study the relations between the set of exact solutions $\mathcal{X}(\mathcal{b})$ (see Definition 2.4) to the control problem (2.3) and sdc-optimal solutions if the number of data m tends to infinity and the observation errors are assumed to get small.

Let be satisfied almost all the assumptions of Theorem 3.17 for a model quadruple $(m_i; Q^{m_i}; \delta_i; z^{(i)})$, $i=1,2,\ldots$, with $m_1 < m_2 < \ldots < m_i < m_{i+1} \ldots \to \infty$. However, instead of $B_0 \cap \mathcal{Z}^* \neq \emptyset$ we postulate $\mathcal{X}_c^* := \mathcal{X}(\mathcal{b}) \cap W_c^\pi \neq \emptyset$. Moreover, the symbols $\mathcal{X}^{(i)}$, $i=1,2,\ldots$, designate here the nonempty sets of minimizers to the problems

$$\underset{x \in W_c^\pi \cap \mathcal{D}}{\text{minimize}} \ \|Q^{m_i}\mathcal{A}x - z^{(i)}\|^2, \quad i=1,2,\ldots, \tag{3.34}$$

respectively

<u>Theorem 3.24:</u>

Under the assumptions stated above, we obtain

$$\emptyset \neq \limsup_{i\to\infty} \mathcal{X}^{(i)} \subseteq \mathcal{X}_c^* \subseteq \mathcal{X}(\mathcal{b}) \ . \ \circ \tag{3.35}$$

<u>Proof:</u>

Let $x^{(i)} \in \mathcal{X}^{(i)}$, $i=1,2,\ldots$, be a sequence of solutions to (3.34). Since $W_c^\pi \cap \mathcal{D}$ is compact in $B_1$, we have a subsequence $x^{(i_j)} \xrightarrow[1]{} x' \in W_c^\pi \cap \mathcal{D}$. In order to prove this theorem, we still have to show that $x' \in \mathcal{X}_c^*$. The inequality

$\|Q^{m_{i_j}}\mathcal{A}x^{(i_j)} - z^{(i_j)}\|^2 \leq \|Q^{m_{i_j}}\mathcal{A}x_c^* - z^{(i_j)}\|^2$ holds for all j and $x_c^* \in \mathcal{X}_c^*$. Thus, the formula (3.35) is valid whenever

$\lim_{j\to\infty} \|Q^{m_{i_j}}\mathcal{A}x^{(i_j)} - z^{(i_j)}\| = \|\mathcal{A}x' - \mathcal{b}\|_2$, but this limiting property has been shown within the proof of Theorem 3.17. #

As we have seen, the required conditions in order to show that optimal solutions to a control problem (2.3) may be approximated arbitrarily well by sdc-optimal solutions are similar to those in the identification case for the convergence of sdi-optimal solutions to exact solutions of (2.1).

## 3.2. Optimal Solutions to Noisy Discretized Problems

### 3.2.1. Model Assumptions

The stability and approximation properties of Sec. 3.1. guarantee that sdi-optimal and sdc-optimal solutions continuously depend on the input data. Furthermore, if the measurement errors tend to become small enough and the data vector dimension m grows to infinity, then the optimal solutions to the noisy semi-discretized inverse problems may approximate solution elements to the inverse problems (2.1) and (2.3) arbitrarily well. However, the optimization problems (3.8) and (3.33) are to be solved. In general, this solution requires an a priori discretization as introduced in Sec. 2.1.. Instead of the Banach space solution element, we now quest for its finite dimensional skeleton. Thus, practical reasons force to replace the semi-discretization model by a full-discretization counterpart, before the computer can begin to do its job. Recall the definitions of discretized inverse problems (dee Definition 2.9), noisy data problems (Definition 2.63) and of the semi-discretization model (Definition 3.1 and Definition 3.20). So we can define the full-discretization model.

Definition 3.25:

Consider a semi-discretization model of an inverse problem and operators $P^n$, $\varphi^n$ and A as introduced in Sec. 2.1.. Then we call (2.1) or (2.3) with an octuple $(m;n;Q^m;P^n;\varphi^n;A;\delta;z)$ of intrinsic model ingredients full-discretization model of the identification or control problem, respectively. The full-discretization model is associated with noisy data discretized inverse problems. $\circ$

Assumption 3.26:

In the sequel, let $\Omega: R^n \to R$ be a continuous functional satisfying the following two conditions.

VIII. $\Omega(x) \geq 0$, for all $x \in R^n$, i.e., $\Omega$ is a nonnegative functional.

IX. The level sets

$$W_c^{\Omega} := \{x \in R^n : \ \Omega(x) \leq c\} \tag{3.36}$$

are, for any real number $c \geq 0$, empty or bounded closed subsets of $R^n$, i.e., $\Omega$ is a stabilizing functional.

Moreover, the operator $\varphi^n: D \subseteq R^n \to \mathcal{D} \subseteq B_1$ with $D := P^n\mathcal{D}$ is assumed to satisfy the conditions

X. $\wp^n x \in B_0 \cap \mathcal{D}$, for all $x \in D$,

and

XI. $P^n \wp^n x = x$, for all $x \in D$.

Finally, assume that the set $\mathcal{X}_{opt}$ of sdi-optimal solutions (see Definition 3.5) contains at least one element $\varkappa_{opt}$ such that the inequality

$$\| Q^m \mathcal{A} \varkappa_{opt} - A P^n \varkappa_{opt} \| \leq h_1 + h_2 \sqrt{\Omega(P^n \varkappa_{opt})} \tag{3.37}$$

is fulfilled for a certain pair of nonnegative numbers $h_1$ and $h_2$. ○

Remark 3.27:

The functional $\Omega$ serves as a finite dimensional analogue of $\Pi$. Any theory of discretized inverse problems has to make assumptions on the behaviour of the discretization error $Q^m \mathcal{A}\varkappa - AP^n\varkappa$. In this text, we propose the inequality (3.37), which allows us to handle errors of approximation growing to infinity with the norm of the approximated Banach space element. The term $h_1 + h_2 \cdot \sqrt{\Omega(P^n \varkappa)}$ is related to expressions arising in the formulation of the generalized discrepancy principle (see GONCHARSKI et al. [162]) also [434,p.106]). The special square root form helps to treat mathematically the approximation error (see the present and the coming chapter). ○

The full-discretization counterpart to the set $\mathcal{Z}_\delta$ (see formula (3.1)) of elements compatible with the semi-discretization model is formed by

$$Z_{\delta,h} := \left\{ x \in D : \|Ax - z\| \leq \delta + h_1 + h_2 \sqrt{\Omega(x)} \right\} . \tag{3.38}$$

Due to Assumption 3.26, the closed set $Z_{\delta,h}$ is never empty. Therefore, it seems to be natural to define optimal solutions as follows:

Definition 3.28:

Any global solution $x_{opt} \in D$ to the optimization problem

$$\begin{array}{l} \text{minimize } \Omega(x) \\ x \in Z_{\delta,h} \end{array} \tag{3.39}$$

is called optimal solution to the noisy data discretized identification problem (for short, di-optimal solution). The set of all di-optimal solutions will be denoted by $X_{opt}$. ○

Definition 3.29:

For given $c \geq 0$, any global solution $x_c \in W_c^{\Omega} \cap D$ to the optimization problem

$$\begin{array}{l} \text{minimize } \| A x - z \|^2 \\ x \in W_c^{\Omega} \cap D \end{array} \tag{3.40}$$

is called optimal solution to the noisy data discretized control problem (for short, dc-optimal solution). The set of all dc-optimal solutions will be denoted by $X_c$. O

In view of Assumption 3.26, the optimization problem (3.39) and provided $W_c^{\Omega} \cap D \neq 0$ (3.40) are well posed on $R^n$ in the sense of Definition 3.11. Theorems concerned with the stability with respect to constraint perturbations may be formulated similar to Theorem 3.16.

### 3.2.2. Asymptotic Properties

Now consider the relations between sdi-optimal and di-optimal solutions as well as between sdc-optimal and dc-optimal solutions, i.e., the asymptotic behaviour of the full-discretization model if the dimension number n tends to infinity. In this context, let denote $(m;n_i;Q^m;P^{n_i};\wp^{n_i};A_i;\delta;z)$, $i=1,2,\ldots$, with $n_1<n_2<\ldots<n_i<n_{i+1}<\ldots\to\infty$ a sequence of octuples associated with full discretization models of a submitted inverse problem (2.1) or (2.3). A functional $\Omega^{(i)}: R^{n_i}\to R$ subject to Assumption 3.26 and a pair $h_1^{(i)}$, $h_2^{(i)}$ of nonnegative real numbers are assumed to correspond to any positive integer i. Furthermore, let us suppose that the inequality

$$\|Q^m\mathcal{A}\varkappa_{opt}-A_iP^{n_i}\varkappa_{opt}\|\leq h_1^{(i)}+h_2^{(i)}\cdot\sqrt{\Omega^{(i)}(P^{n_i}\varkappa_{opt})} \tag{3.41}$$

holds for a certain element $\varkappa_{opt}\in\mathcal{X}_{opt}$ and all $i=1,2,\ldots$ . Then in the identification case we have:

Theorem 3.30:
Let $\{x^{(i)}\in R^{n_i}\}_{i=1}^{\infty}$ denote a sequence of di-optimal solutions according to

$$\underset{x\in Z^{(i)}}{\text{minimize}}\ \Omega^{(i)}(x) \tag{3.42}$$

with

$$Z^{(i)}:=\{x\in P^{n_i}\mathcal{D}:\|A_ix-z\|\leq\delta+h_1^{(i)}+h_2^{(i)}\sqrt{\Omega^{(i)}(x)}\}. \tag{3.43}$$

Furthermore, let $\{\varkappa^{(i)}=\wp^{n_i}x^{(i)}\}_{i=1}^{\infty}\subseteq B_0\cap\mathcal{D}$ be the associated sequence of elements projected back to the Banach space $B_1$. Provided that, for all $\varkappa\in\mathcal{D}$ and all integers i,

$$\|Q^m\mathcal{A}\wp^{n_i}P^{n_i}\varkappa-A_iP^{n_i}\varkappa\|\leq\psi_1^{(i)}+\psi_2^{(i)}\cdot\|\wp^{n_i}P^{n_i}\varkappa\|_1 \tag{3.44}$$

and

$$\Pi(\wp^{n_i}P^{n_i}\varkappa)\leq(1+\psi_3^{(i)})\cdot\Omega^{(i)}(P^{n_i}\varkappa), \tag{3.45}$$

moreover, for all $\varkappa\in B_0\cap\mathcal{D}$,

$$\limsup_{i\to\infty}\Omega^{(i)}(P^{n_i}\varkappa)\leq\Pi(\varkappa) \tag{3.46}$$

and finally $h_1^{(i)}$, $h_2^{(i)}$, $\psi_1^{(i)}$, $\psi_2^{(i)}$ and $\psi_3^{(i)}$ are null-sequences, then the sequence $\{\varkappa^{(i)}\}_{i=1}^{\infty}$ is relatively compact in $B_1$ and any point of accumulation of this sequence lies in the set $\mathcal{X}_{opt}$ of sdi-optimal solutions. ○

Proof:

Suppose that $\varkappa_{opt} \in \mathcal{X}_{opt}$ satisfies the inequality (3.41). By the triangle inequality one obtains $\| A_i P^{n_i} \varkappa_{opt} - z \| \leq \| A_i P^{n_i} \varkappa_{opt} - Q^m \mathcal{A} \varkappa_{opt} \| + \| Q^m \mathcal{A} \varkappa_{opt} - z \| \leq \delta + h_1^{(i)} + h_2^{(i)} \sqrt{\Omega^{(i)}(P^{n_i} \varkappa_{opt})}$. Thus, $P^{n_i} \varkappa_{opt} \in Z^{(i)}$.

From (3.46) and $\Omega^{(i)}(x^{(i)}) \leq \Omega^{(i)}(P^{n_i} \varkappa_{opt})$ we get $\limsup_{i \to \infty} \Omega^{(i)}(x^{(i)}) \leq \Pi(\varkappa_{opt})$ and with $\varkappa^{(i)} = \mathcal{P}^{n_i} P^{n_i} \varkappa^{(i)} = \mathcal{P}^{n_i} x^{(i)}$, in view of (3.45), $\limsup_{i \to \infty} \Pi(\varkappa^{(i)}) \leq \Pi(\varkappa_{opt})$. Since $\Pi$ is a stabilizing functional (see Assumption 3.3), the sequence $\{\varkappa^{(i)}\}_{i=1}^{\infty}$ is relatively compact. If we still show that any point of accumulation $\varkappa'$ of this sequence lies in the set $Z_\delta$, then $\Pi(\varkappa') \leq \Pi(\varkappa_{opt})$, $\varkappa' \in \mathcal{X}_{opt}$, and the theorem proved to be correct. Now,

$\| Q^m \mathcal{A} \varkappa^{(i)} - z \| \leq \| Q^m \mathcal{A} \mathcal{P}^{n_i} x^{(i)} - A_i x^{(i)} \| + \| A_i x^{(i)} - z \| \leq \psi_1^{(i)} + \psi_2^{(i)} \cdot \| \varkappa^{(i)} \|_1 + \delta + h_1^{(i)} + h_2^{(i)} \sqrt{\Omega^{(i)}(x^{(i)})}$. Owing to formula (3.4) this implies

$$\| Q^m \mathcal{A} \varkappa^{(i)} - z \| \leq \delta + \psi_1^{(i)} + h_1^{(i)} + \psi_2^{(i)} \cdot \frac{\sqrt{\Pi(\varkappa^{(i)})}}{c} + h_2^{(i)} \sqrt{\Omega^{(i)}(x^{(i)})} \leq$$

$$\leq \delta + \psi_1^{(i)} + h_1^{(i)} + \left( \psi_2^{(i)} \sqrt{\frac{1+\psi_3^{(i)}}{c}} + h_2^{(i)} \right) \cdot \sqrt{\Omega^{(i)}(x^{(i)})}$$

. Since $\{\Omega^{(i)}(x^{(i)})\}_{i=1}^{\infty}$ is a bounded sequence of real numbers, we can derive, for any accumulation point $\varkappa' \in B_0 \cap \mathcal{D}$ of the sequence $\{\varkappa^{(i)}\}_{i=1}^{\infty}$, $\| Q^m \mathcal{A} \varkappa' - z \| \leq \delta$ and hence $\varkappa' \in Z_\delta$. #

Remark 3.31:

The theorem stated above gives sufficient conditions for the asymptotic approximation of sdi-optimal solutions by di-optimal solutions. Note that the requirements in the formulae (3.44) through (3.46) do not affect the arising operators $\mathcal{A}$, A, $P^n$, $\mathcal{P}^n$, $Q^m$ and functionals $\Pi$ and $\Omega$ itself, but only compounds of these operators and functionals in a few combinations. An adaptive choice of the not necessarily linear back projector $\mathcal{P}^n$ can help fulfil these requirements. The example of Remark 3.33 will illustrate the assumptions of Theorem 3.10 by means of a rather simpler case. A schematic survey of the arising spaces, sets and operators is given by the following figure. ○

Figure 3.32:

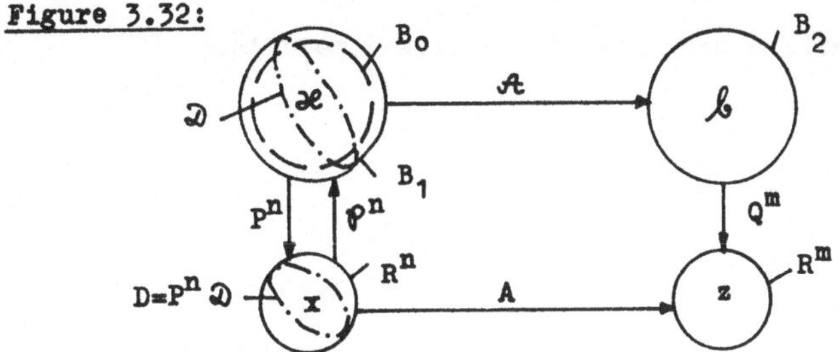


Remark 3.33:

Referring back to the situation of paragraph c. of Remark 3.10, we consider an unconstrained linear identification problem (3.1) with compact operator $\mathcal{A}: H_1 \to H_1$ in a Hilbert space. For the discretization, we use similar to Sec. 2.2.2.:

$P^n\varkappa := x = ((\varkappa, u^{(1)})_1, \ldots, (\varkappa, u^{(n)})_1)^T$, $Q^m\mathscr{b} := ((\mathscr{b}, u^{(1)})_1, \ldots, (\mathscr{b}, u^{(m)})_1)^T$,

$A := ((\mathcal{A}u^{(i)}, u^{(j)})_1 \;)^{j=1(1)m}_{i=1(1)n}$, $\mathcal{P}^n x := \sum_{i=1}^{n} x_i \cdot u^{(i)}$ and $\Omega(x) := \sum_{i=1}^{n} \lambda_i x_i^2$.

Obviously, for all $x \in R^n$, $\|Q^m\mathcal{A}\mathcal{P}^n x - Ax\| = 0$. Thus, (3.44) holds with $\Psi_1 = \Psi_2 = 0$. Furthermore, $\Pi(\mathcal{P}^n P^n \varkappa) = \Omega(P^n\varkappa)$. The inequality (3.45) with $\Psi_3 = 0$ is proved to be valid. Since $\Omega(P^n\varkappa) = \sum_{i=1}^{n} \lambda_i (\varkappa, u^{(i)})_1^2$ tends to $\Pi(\varkappa) = \sum_{i=1}^{\infty} \lambda_i \cdot (\varkappa, u^{(i)})_1^2$ as $n \to \infty$, the condition (3.46) also holds. The approximation error norm square $\|Q^m\mathcal{A}\varkappa - AP^n\varkappa\|^2 =$ $\sum_{j=1}^{m} ( \sum_{i=n+1}^{\infty} (\mathcal{A}u^{(i)}, u^{(j)})_1 (\varkappa, u^{(i)})_1)^2 \leq \sum_{j=1}^{m} \{ ( \sum_{i=n+1}^{\infty} (\mathcal{A}u^{(i)}, u^{(j)})_1^2 ) \cdot$ $( \sum_{i=n+1}^{\infty} (\varkappa, u^{(i)})_1^2 ) \}$ converges to zero as $n \to \infty$ owing to $\|\mathcal{A}^* u^{(j)}\|_1^2 =$ $\sum_{i=1}^{\infty} (\mathcal{A}u^{(i)}, u^{(j)})_1^2 < \infty$. Inequality (3.41) characterizes for at least one element $\varkappa = \varkappa_{opt} \in \mathcal{X}_{opt}$ a requirement on the decreasing behaviour of the null-sequence $\sum_{i=n+1}^{\infty} (\varkappa, u^{(i)})_1^2$ as n increases without bounds. ○

A further example to Theorem 3.30 concerning the Banach space case introduced in paragraph b. of Remark 3.10 with $\Pi$ and $\Omega$ according to the formulae (2.94) and (2.95) is given in [209, p.36]. Now let us turn to the control case. As a counterpart to (3.41) we have to suppose here

$$\Omega^{(i)}(P^{n_i}\varkappa_c) \leq \Pi(\varkappa_c) + h_3^{(i)}, \quad h_3^{(i)} \geq 0, \tag{3.47}$$

for at least one sdc-optimal solution $\varkappa_c \in \mathcal{X}_c$, i=1,2,... .

Theorem 3.34:

Let $\{x^{(i)} \in R^{n_i}\}_{i=1}^{\infty}$ denote a sequence of dc-optimal solutions according to

$$\underset{x \in W^{\Omega^{(i)}}_{c+h_3^{(i)}} \cap D}{\text{minimize}} \|A_i x - z\|^2 \tag{3.48}$$

and $\{\varkappa^{(i)} = \rho^{n_i} x^{(i)}\}_{i=1}^{\infty} \subseteq B_0 \cap \mathcal{D}$ the associated sequence of elements projected back to the Banach space $B_1$. Provided that, for all $\varkappa \in B_0 \cap \mathcal{D}$, the inequality

$$\|Q^m \mathcal{A}\varkappa - A_1 P^n \varkappa\| \leq \psi_1^{(i)} + \psi_2^{(i)} \cdot \|\varkappa\|_1 \tag{3.49}$$

and, for all $\varkappa \in \mathcal{D}$, the inequality (3.45) hold and $h_3^{(i)}$, $\psi_1^{(i)}$, $\psi_2^{(i)}$, $\psi_3^{(i)}$ are null-sequences, $\{\varkappa^{(i)}\}_{i=1}^{\infty}$ is relatively compact in $B_1$ and any point of accumulation lies in the set $\mathcal{X}_c$ of sdc-optimal solutions. ○

Proof:

Suppose that $\varkappa_c \in \mathcal{X}_c$ satisfies (3.47). Obviously, $\Omega^{(i)}(P^n \varkappa_c) \leq \Pi(\varkappa_c) + h_3^{(i)}$, i.e., $P^n \varkappa_c \in W^{\Omega^{(i)}}_{c+h_3^{(i)}} \cap D$ for all i. Consequently, we obtain $\|A_1 x^{(i)} - z\|^2 \leq \|A_1 P^n \varkappa_c - z\|^2$ for all i and due to (3.45) $\Pi(\varkappa^{(i)}) \leq (1 + \psi_3^{(i)}) \Omega^{(i)}(x^{(i)})$. This implies with $\limsup_{i \to \infty} \Pi(\varkappa^{(i)}) \leq c$ that $\{\varkappa^{(i)}\}_{i=1}^{\infty}$ is relatively compact in $B_1$. Any point of accumulation $\varkappa' \in B_0 \cap \mathcal{D}$ of this sequence belongs to the set $W_c^{\Pi} \cap \mathcal{D}$. In order to prove the theorem, it now suffices to show that the inequality $\|Q^m \mathcal{A}\varkappa' - z\|^2 \leq \|Q^m \mathcal{A}\varkappa_c - z\|^2$ holds. However, applying the triangle inequality twice, we obtain $\|Q^m \mathcal{A}\varkappa^{(i)} - z\| \leq \|Q^m \mathcal{A}\varkappa^{(i)} - A_1 x^{(i)}\| + \|A_1 x^{(i)} - z\| \leq \psi_1^{(i)} + \psi_2^{(i)} \cdot \|\varkappa^{(i)}\|_1 + \|A_1 x^{(i)} - z\| \leq \psi_1^{(i)} + \psi_2^{(i)} \cdot \|\varkappa^{(i)}\|_1 + \|A_1 P^n \varkappa_c - z\| \leq \psi_1^{(i)} + \psi_2^{(i)} \cdot \|\varkappa^{(i)}\|_1 + \|A_1 P^n \varkappa_c - Q^m \mathcal{A}\varkappa_c\| + \|Q^m \mathcal{A}\varkappa_c - z\| \leq 2 \cdot \psi_1^{(i)} + \psi_2^{(i)}(\|\varkappa^{(i)}\|_1 + \|\varkappa_c\|_1) + \|Q^m \mathcal{A}\varkappa_c - z\|$. Following the arguments of the proof of Theorem 3.30, this implies the needed inequality. A requirement of the form (3.46) seems to be superfluous in the control case, whereas (3.49) is stronger than (3.44). #

## 3.3. Uncertainty of Approximate Solutions

In the general optimization approach to the approximate solution of inverse problems introduced in the Secs. 3.1. and 3.2., the ill-posed identification and control problems (see Definitions 2.5 and 2.6) are replaced by well-posed optimization problems (3.8), (3.33), (3.39) and (3.40). The quality of optimal solutions thus obtained depends upon measurement errors, upon the discretization model, the dimension numbers m and n and associated operators, upon objective (sets $\mathfrak{D}$ and D) and subjective (functionals $\Pi$ and $\Omega$) a priori information.

Note that the subjective a priori information applies to identification and control in a different way. On the one hand, in the identification case, the evaluation functional will be minimized over the set of elements (see (3.8), (3.39)), which are compatible with the measurement errors and, if appropriate, with the discretization errors. On the other hand, an upper bound c is prescribed on the values of $\Pi(\varkappa)$ and $\Omega(x)$ in the control case (see (3.33), (3.40)). Then, optimal solutions have the form of constrained least squares solutions. The prescribed value c can be interpreted as the admissible minimum level of credibility for a solution element $\varkappa$ or x. For both classes of inverse problems, the quality of approximate solutions will also depend on the more or less appropriate choice of the constants $\delta$, $h_1$, $h_2$ and c.

If our consideration is confined to the practically most important full-discretization model with a fixed intrinsic octuple (see Definition 3.25), then a frequently considerable error (uncertainty) in the approximate solution will be inevitable. At the end of this Chapter 3 we are going to study this uncertainty in the sense of a worst case error analysis (cf. [213], also LINZ [279] and FRANKLIN [131]).for the identification case. Simultaneously, close connections between the formulations (3.39) and (3.40) of identification and control problems become evident.

In this brief study, we are interested in reconstructing an element $\varkappa^* \in \mathfrak{D}$ satisfying (3.1). It is supposed that this element represents a real material or process characteristic. Let be given a real discretization error bound h with

$$\| Q^m \mathcal{A} \varkappa^* - A P^n \varkappa^* \| \leq h \ . \tag{3.50}$$

Our aim is to find the skeleton solution $x^* := P^n \varkappa^*$. Therefore, we search for procedures $\hat{x} = \hat{x}(z)$, which yield approximate solutions $\hat{x}$ with small deviations $\| \hat{x} - x^* \|$ to any given data vector z. If we considered a stochastic noisy data problem, the random character of the measurement error vector y would suggest designating $\hat{x}(z)$ as an estimator and the vector $\hat{x}$ as an estimate of $x^*$. We will use these denotations here, although the stochastic character of y is not needed for the study. In addition to the above introduced level sets $W_c^{\Omega}$ let

$$W_*^{\Omega} := \{x \in R^n : \Omega(x) \leq \Omega(x^*)\} \quad . \tag{3.51}$$

denote the particular level set corresponding to the exact solution.

The following worst case error analysis is based on the modulus functional subject to the possibly multi-valued inverse mapping $A^{(-1)}$ to A

$$\omega_{A^{(-1)}}(s,S) := \sup_{\substack{x^{(1)},\, x^{(2)} \in S \\ \|Ax^{(1)}-Ax^{(2)}\| \leq s}} \|x^{(1)} - x^{(2)}\| \quad . \tag{3.52}$$

All assertions of Lemma 3.35 are well-known.

<u>Lemma 3.35:</u>

The modulus functional $\omega_{A^{(-1)}}(s,S)$ is, for fixed $S \subseteq D$, monotonically nondecreasing with respect to $s \geq 0$, where $\omega_{A^{(-1)}}(s,S) = \infty$ need not be excluded, in general. If $S_1 \subseteq S_2 \subseteq D$, then $\omega_{A^{(-1)}}(s,S_1) \leq \omega_{A^{(-1)}}(s,S_2)$ for fixed $s \geq 0$. We have $\omega_{A^{(-1)}}(s,S) \leq \mathrm{diam}(S) := \sup_{x^{(1)},x^{(2)} \in S} \|x^{(1)}-x^{(2)}\| < \infty$, for all $s \geq 0$, whenever S is a bounded subset of $R^n$. Moreover, if A is injective over D, then $\omega_{A^{(-1)}}(s,S)$ is the modulus of continuity to the inverse operator $A^{-1}: AS \rightarrow S$. A continuous inverse $A^{-1}$ is characterized by the limit $\lim_{s \to 0} \omega_{A^{(-1)}}(s,S) = 0$ . ○

As previously done in the control case, we introduce a lower credibility bound c here, too.

<u>Lemma 3.36:</u>

Let $\hat{x} \in W_c^{\Omega} \cap D$ be an estimate of $x^*$. Then,

$$\|\hat{x}-x^*\| \leq \bar{\omega}(\hat{x}) := \omega_{A^{(-1)}}(\delta+h+\|A\hat{x}-z\|, (W_c^{\Omega} \cup W_*^{\Omega}) \cap D). \tag{3.53}$$

The functional $\bar{\omega}(\hat{x})$ represents an uncertainty measure in the worst case sense. ○

<u>Proof:</u>

With (3.50) and (2.97) the triangle inequality yields $\|Ax^*-z\| \leq \|Ax^*-Q^m \mathcal{R} \varkappa^*\| + \|Q^m \mathcal{R} \varkappa^* - z\| \leq \delta + h$ and $\|A\hat{x}-Ax^*\| \leq \|A\hat{x}-z\| + \|z-Ax^*\|$. This implies the inequality (3.53). #

Immediately from this lemma we can derive the estimator $\hat{x}(z) \in W_c^{\Omega} \cap D$ for $x^*$ with minimum uncertainty.

Corollary 3.37:

The uncertainty measure $\bar{\omega}(\hat{x})$ defined by formula (3.53) attains its minimum over $\hat{x} \in W_c^{\Omega} \cap D$ if $x := x_c$ is a minimizer of the constrained least squares problem (3.40), i.e., a dc-optimal solution. ○

Note that any solution $x_c$ to problem (3.40) is a quasisolution in the sense of IVANOV [222-23] (see also [434, Chap.2] and TICHATSCHKE/HOFMANN [426]). Its general representation possesses the form

$$x_c = A^{(-1)} \mathbb{P}_{A(W_c^{\Omega} \cap D)} z, \tag{3.54}$$

where $\mathbb{P}_{A(W_c^{\Omega} \cap D)} z \in A(W_c^{\Omega} \cap D)$ denotes the not necessarily unique Euclidean norm projection element of a vector z into the compact set $A(W_c^{\Omega} \cap D)$ (range of the operator A restricted to the domain $W_c^{\Omega} \cap D$). The projection is defined as

$$\| \mathbb{P}_{A(W_c^{\Omega} \cap D)} z - z \| = \min_{\tilde{z} \in A(W_c^{\Omega} \cap D)} \| \tilde{z} - z \| . \tag{3.55}$$

If (3.55) determines a unique projection vector, then the dc-optimal solution is also unique whenever A is injective over $W_c^{\Omega} \cap D$.

Corollary 3.38:

For $\bar{\omega}(x_c)$, one can obtain majorant functions of c as follows:

$$\|x_c - x^*\| \le \bar{\omega}(x_c) := \omega_{A^{(-1)}}(\delta + h + \|Ax_c - z\|, (W_c^{\Omega} \cup W_*^{\Omega}) \cap D) \tag{3.56}$$

$$\le \begin{cases} \bar{\omega}_1(c) := \omega_{A^{(-1)}}(\delta + h + \|Ax_c - z\|, W_*^{\Omega} \cap D) & \text{if } \min_{x \in D} \Omega(x) \le c \le \Omega(x^*), \\ \bar{\omega}_2(c) := \omega_{A^{(-1)}}(2(\delta + h), W_c^{\Omega} \cap D) & \text{if } \Omega(x^*) \le c < \infty. \end{cases}$$

Moreover, we have $\Omega(x_{c_1}) \le \Omega(x_{c_2})$ and $\|Ax_{c_1} - z\| \ge \|Ax_{c_2} - z\|$ if $\min_{x \in D} \Omega(x) \le c_1 < c_2 < \infty$ . ○

Considering Lemma 3.35 it may be easily derived that $\bar{\omega}_1(c)$ is a monotonically nonincreasing function of c in the interval $\min_{x \in D} \Omega(x) \le c \le \Omega(x^*)$ whereas $\bar{\omega}_2(c)$ is monotonically nondecreasing with respect to $\Omega(x^*) \le c < \infty$. This would suggest substituting $c := \Omega(x^*)$ in order to achieve an appropriate trade-off between sets $W_c^{\Omega} \cap D$, which become augmented if c gets too large, and increasing residual norms $\|Ax_c - z\|$ if c is chosen too small. The associated worst case error can be written by formula

$$\| x_{\Omega(x^*)} - x^* \| \le \omega_{A^{(-1)}}(2(\delta + h), W_*^{\Omega} \cap D) . \tag{3.57}$$

That means, if $\Omega(x^*)$ were a known value, a certain dc-optimal solution would satisfactorily help to solve identification problems.

Unfortunately, in almost all cases $\Omega(x^*)$ fails to be available, whereas there exist real chances of obtaining values for the sum $\delta+h$. Then, di-optimal solutions are computable. As the coming theorem indicates, they yield approximate solutions, the uncertainty of which corresponds to that of formula (3.57).

Theorem 3.39:

Let

$$Z_{\delta,h} := \{x \in D : \|A x - z\| \leq \delta + h\} \tag{3.58}$$

denote the set of compatible elements and $x_{opt}$ a minimizer of

$$\underset{x \in Z_{\delta,h}}{\text{minimize}} \ \Omega(x) \ , \tag{3.59}$$

i.e., a particular di-optimal solution associated with the model of this section. Then,

$$\|x_{opt} - x^*\| \leq \omega_{A^{(-1)}}(2(\delta+h),\ W_*^\Omega \cap D) \ . \ \circ \tag{3.60}$$

Proof:

As we already have proven, $\|Ax^*-z\| \leq \delta+h$. Therefore, $x^* \in Z_{\delta,h}$, $\Omega(x_{opt}) \leq \Omega(x^*)$ and consequently $x_{opt} \in W_*^\Omega \cap D$. In view of $\|Ax_{opt}-Ax^*\| \leq \|Ax_{opt}-z\| + \|z-Ax^*\| \leq 2(\delta+h)$, this implies the formula (3.60). #

The modulus functional is also suited for evaluating the content of information of a discretized identification problem in dependence of the operator A, the error bounds $\delta$ and h, the set D and the functional $\Omega$ related to the really submitted vector $x^*$. Provided the optimal estimate $x_{opt}$ is used, then the intrinsic value $\omega_{A^{(-1)}}(2(\delta+h),\ W_*^\Omega \cap D)$ measures the remaining uncertainty. With diminishing error bounds $\delta$ and h, the vector $x_{opt}$ tends to approach $x^*$. Furthermore, the intrinsic value also helps compare the aptitude of operator A and the degree of a priori information. If the subjective a priori information is entirely missing, an arbitrarily chosen vector $x_{\delta,h} \in Z_{\delta,h}$ satisfies the inequality

$$\|x_{\delta,h} - x^*\| \leq \omega_{A^{(-1)}}(2(\delta+h),\ D) \ . \tag{3.61}$$

Thus, the difference $\omega_{A^{(-1)}}(2(\delta+h),\ D) - \omega_{A^{(-1)}}(2(\delta+h),\ W_*^\Omega \cap D)$ expresses the gain of precision achieved by the additional use of the functional $\Omega$.

Until now we have considered the properties of optimal solutions to noisy data problems without asking how to solve the problems (3.39) and (3.40) numerically. The following chapter is inter alia devoted to this aspect.

# 4. Regularization of Deterministic Discretized Inverse Problems

## 4.1. Constrained Tikhonov Regularization

### 4.1.1. Regularized Solutions

In this chapter, we continue to treat noisy data problems of identification and control under deterministic assumptions on the error of observation. Within Chapter 3, auxiliary problems of minimization type (3.39) and (3.40) have been constructed, the solutions of which continuously depend on the input data. Now we present a family of well-posed optimization problems that represent stable neighbouring problems for the discretized inverse problem under consideration. The idea of dealing with this family of problems goes back to TIKHONOV [428-29]. Therefore, the method considered below in a particular fashion is called Tikhonov regularization method.

From the extensive literature regarding this method we only mention the monographs of TIKHONOV/ARSENIN [434], FRIEDRICH et al. [138], and GROETSCH [177] as well as the reports of BJÖRCK/ELDÉN [33] and WAHBA [470]. We shall term the method constrained Tikhonov regularization if a set $D \neq R^n$ reflecting objective a priori information is taken into account. In the sequel, essential properties of regularized solutions are summarized and the interrelations between the Tikhonov regularization method, di-optimal and dc-optimal solutions shall become clear.

Definition 4.1:

Let

$$F(x,\alpha) := \|A\,x - z\|^2 + \alpha\cdot\Omega(x)\,, \quad x\in D,\ \alpha>0 \tag{4.1}$$

be Tikhonov's functional for a discretized inverse problem. It is defined as a sum of the Euclidean residual norm square and a term in which the positive regularization parameter $\alpha$ is multiplied by the stabilizing functional $\Omega$ satisfying the Assumption 3.26. We say that any global solution of the optimization problem

$$\underset{x\in D}{\text{minimize}}\ \ F(x,\alpha)\,, \tag{4.2}$$

for given regularization parameter $\alpha>0$, is a regularized solution and we denote it by $x_{(\alpha)}$. The parentheses enclosing $\alpha$ at the lower right-hand corner of $x$ are written in order to distinguish dc-optimal solutions and regularized solutions. Given $\alpha>0$, we denote by $X_{(\alpha)}$ the set of all regularized solutions $x_{(\alpha)}$. Due to the Euclidean norm square form of the first term in the functional (4.1) the presented fashion of Tikhonov regularization is sometimes called a least squares regularization. ○

Remark 4.2:

In order to ensure that (4.2) has a unique minimizer $x_{(\alpha)}$, strong assumptions about A, D and $\Omega$ are needed. Provided $\Omega$ is strictly convex, i.e.,

$$\Omega\left(\frac{x^{(1)}+x^{(2)}}{2}\right)<\frac{1}{2}\cdot\left(\Omega(x^{(1)})+\Omega(x^{(2)})\right), \text{ for all } x^{(1)},x^{(2)}\in D, \quad (4.3)$$

the linearity of A is sufficient for the uniqueness of a minimizer of (4.2) in D. Uniqueness of regularized solutions in the case of nonlinear problems is even associated with a seldom available convexity condition of the form

$$\left\|A\left(\frac{x^{(1)}+x^{(2)}}{2}\right)-z\right\|^2\leq\frac{1}{2}\cdot\left(\|Ax^{(1)}-z\|^2+\|Ax^{(2)}-z\|^2\right),$$
$$\text{for all } x^{(1)},x^{(2)}\in D \text{ and } z\in R^m \; . \circ \quad (4.4)$$

Lemma 4.3:

Due to the previously stated assumptions about D and $\Omega$, the sets $X_{(\alpha)}$ of regularized solutions $x_{(\alpha)}$ are nonempty, closed and bounded subsets of D for all $\alpha>0$. The sets

$$X\begin{Bmatrix}\alpha_2\\ \alpha_1\end{Bmatrix}:=\bigcup_{0<\alpha_1\leq\alpha\leq\alpha_2}X_{(\alpha)}\;,\; X\begin{Bmatrix}\infty\\ \alpha_1\end{Bmatrix}:=\bigcup_{0<\alpha_1\leq\alpha<\infty}X_{(\alpha)} \quad \text{and}$$

$X_{(\infty)}:=\{x\in D:\ \Omega(x)=\Omega_{min},\ \Omega_{min}:=\min_{x\in D}\Omega(x)\}$ possess the same properties. Moreover, the optimization problem (4.2) is well-posed on $R^n$ for all $\alpha>0$. $\circ$

Definition 4.4:

Any global solution $x_{ls}\in D$ of the optimization problem

$$\underset{x\in D}{\text{minimize}}\ \|Ax-z\|^2 \quad (4.5)$$

is termed a least squares solution. Analogously, we denote by $X_{ls}$ the set of all least squares solutions. $\circ$

Remark 4.5:

If D is unbounded, then the closed set $X_{ls}$ of least squares solutions $x_{ls}$ may also be unbounded or empty. Thus, problem (4.5) may be ill-posed. However, the infimum value $\mu:=\inf_{x\in D}\|Ax-z\|\geq 0$ exists. The minimizer $x_{ls}$ is unique,
for instance, if A is linear and injective over D. $\circ$

Remark 4.6:

Note that the optimization problems (4.2) and (4.5) possess a convex set of feasible solutions D, whereas in (3.39) and (3.40) the objective functionals $\Omega(x)$ and $\|Ax-z\|^2$ are to be minimized subject to a generally nonconvex set $Z_{\delta,h}$ or $W_c^{\Omega}\cap D$, respectively.

Hence, the numerical computation of regularized solutions and least squares solutions seems to be easier than the direct computational solution of the problems (3.39) and (3.40). This is due to the fact that direct numerical methods for handling nonconvex or even nonconnected sets of feasible solutions are not advanced. Therefore, the Tikhonov regularization method is suited for establishing auxiliary problems according to di-optimal and dc-optimal solutions. In the sequel, it will be shown that minimizers $x_{opt}$ and $x_c$ of (3.39) and (3.40) either coincide with or may be well-approximated by regularized solutions for appropriate values of $\alpha$ . ○

By definition of regularized solutions we can derive:

Lemma 4.7:

Let $\{x_{(\alpha_i)}\}_{i=1}^{\infty} \subseteq D$ denote an infinite sequence of regularized solutions with positive regularization parameters $\alpha_i$. If $\alpha_i \geq \underline{\alpha} > 0$, $i=1,2,\ldots$, then this sequence involves a convergent subsequence. If in addition $\alpha_i \to \alpha > 0$ and $x_{(\alpha_i)} \to x^{(1)} \in D$ as $i \to \infty$, then $x^{(1)} \in X_{(\alpha)}$.

If otherwise $\alpha_i \to \infty$ and $x_{(\alpha_i)} \to x^{(2)} \in D$, then $x^{(2)} \in X_{(\infty)}$. Furthermore, provided $X_{ls} \neq \emptyset$, $\alpha_i \to 0$ and $x_{(\alpha_i)} \to x^{(3)} \in D$, then we have $x^{(3)} \in X_{ls}^{\Omega} := \{x \in X_{ls} : \Omega(x) = \min_{\tilde{x} \in X_{ls}} \Omega(\tilde{x})\}$ . On the other hand, whenever $X_{ls} = \emptyset$ , then any sequence $\{x_{(\alpha_i)}\}_{i=1}^{\infty} \subseteq D$ with $\alpha_i \to 0$ satisfies the conditions $\Omega(x_{(\alpha_i)}) \to \infty$ and consequently $\|x_{(\alpha_i)}\| \to \infty$ as $i \to \infty$. ○

We now present some properties regarding limits and the monotonicity of occurring functions. These properties are important for characterizing regularized solutions and their relations to optimal solutions concerning the full-discretization model. The properties are either mentioned and proven in [434] and [438] or can be derived easily. Consider a scale $\langle x_{(\alpha)}, 0 < \alpha < \infty \rangle$ of regularized solutions, i.e., to any regularization parameter $\alpha > 0$ a certain minimizer $x_{(\alpha)}$ of problem (4.2) has been selected. We are going to use the following notation in the sequel: $s_{res}(\alpha) := \|Ax_{(\alpha)} - z\|$, $s_{\Omega}(\alpha) := \Omega(x_{(\alpha)})$ and $s_F(\alpha) := F(x_{(\alpha)}, \alpha) = s_{res}^2(\alpha) + \alpha \cdot s_{\Omega}(\alpha)$. In this chapter, we also simplify the notation of problem (3.39) subject to (3.38) by setting $\vartheta := \delta + h_1$ and $\chi := h_2$. Recalling the Definitions 3.28 and 3.29, we repeat here the formulation of those optimization problems that define $x_{opt}$ and $x_c$. Thus, $x_{opt}$ is a minimizer of

$$\begin{array}{l}\text{minimize } \Omega(x) \\ x\in D \\ \|Ax-z\|\leq\vartheta+\chi\sqrt{\Omega(x)}\end{array} \quad , \quad \vartheta,\chi \geq 0 \;. \tag{4.6}$$

On the other hand, $x_c$ is a minimizer of

$$\begin{array}{l}\text{minimize } \|Ax-z\|^2 \\ x\in D \\ \Omega(x)\leq c\end{array} \quad . \tag{4.7}$$

For the treatment of (4.6), in addition, we introduce the function

$$s_{dif}(\alpha):=\|Ax_{(\alpha)}-z\|-\chi\cdot\sqrt{\Omega(x_{(\alpha)})}=s_{res}(\alpha)-\chi\cdot\sqrt{s_{\Omega}(\alpha)} \;.$$

Lemma 4.8:

The function $s_F(\alpha)$ is continuous; $s_F(\alpha)$, $s_{res}(\alpha)$ and $s_{dif}(\alpha)$ are monotonically nondecreasing functions with respect to $\alpha>0$, $s_{\Omega}(\alpha)$ is a monotonically nonincreasing function with respect to $\alpha>0$. Limit conditions hold as follows:

$$\lim_{\alpha\to\alpha_0-0} s_{res}(\alpha)=\min_{x\in X_{(\alpha_0)}}\|Ax-z\|\leq\max_{x\in X_{(\alpha_0)}}\|Ax-z\|=\lim_{\alpha\to\alpha_0+0} s_{res}(\alpha)\;,$$

$$\lim_{\alpha\to\alpha_0+0} s_{\Omega}(\alpha)=\min_{x\in X_{(\alpha_0)}}\Omega(x)\leq\max_{x\in X_{(\alpha_0)}}\Omega(x)=\lim_{\alpha\to\alpha_0-0} s_{\Omega}(\alpha)\;, \tag{4.8}$$

$$\lim_{\alpha\to\alpha_0-0} s_{dif}(\alpha)=\min_{x\in X_{(\alpha_0)}}\left\{\|Ax-z\|-\chi\sqrt{\Omega(x)}\right\}\leq\max_{x\in X_{(\alpha_0)}}\left\{\|Ax-z\|-\chi\sqrt{\Omega(x)}\right\}=\lim_{\alpha\to\alpha_0+0} s_{dif}(\alpha)\;,\quad \alpha_0>0;$$

$$\lim_{\alpha\to 0+0}\sqrt{s_F(\alpha)}=\lim_{\alpha\to 0+0}s_{res}(\alpha)=\mu\;,$$

$$\lim_{\alpha\to 0+0} s_{\Omega}(\alpha)=\begin{cases}\infty & \text{if } X_{ls}=\emptyset \\ \min\limits_{x\in X_{ls}}\Omega(x) & \text{if } X_{ls}\neq\emptyset\end{cases}\;, \tag{4.9}$$

$$\lim_{\alpha\to 0+0} s_{dif}(\alpha)=\begin{cases}-\infty & \text{if } X_{ls}=\emptyset\;,\ \chi>0 \\ \mu:=\inf\limits_{x\in D}\|Ax-z\| & \text{if } X_{ls}=\emptyset\;,\ \chi=0 \\ \tau_0:=\mu-\chi\cdot\sqrt{\min\limits_{x\in X_{ls}}\Omega(x)} & \text{if } X_{ls}\neq\emptyset\end{cases}\;;$$

$$\lim_{\alpha\to\infty}\sqrt{s_F(\alpha)}=\lim_{\alpha\to\infty}s_{res}(\alpha)=\min_{x\in X_{(\infty)}}\|Ax-z\|\;,$$

$$\lim_{\alpha\to\infty}s_{\Omega}(\alpha)=\Omega_{min}:=\min_{x\in D}\Omega(x)\;, \tag{4.10}$$

$$\lim_{\alpha\to\infty}s_{dif}(\alpha)=\tau_{\infty}:=\min_{x\in X_{(\infty)}}\|Ax-z\|-\chi\cdot\sqrt{\Omega_{min}} \quad\circ$$

Assumption 4.9:
We assume $\vartheta, \chi$ and c to be given so that the optimization problems (4.6) and (4.7) have nonempty sets of feasible solutions. This implies $\Omega_{min} := \min_{x \in D} \Omega(x) \leq c$ and $\min_{x \in D} \{ \|Ax-z\| - \chi\sqrt{\Omega(x)} \} \leq \vartheta$. Moreover, let the functional $\Omega$ be quasimonotonous over D (see [434, 2nd. Russian ed., p.66]), i.e., any local minimizer of the problem

$$\underset{x \in D}{\text{minimize}}\ \Omega(x) \tag{4.11}$$

is a global minimizer $x_{(\infty)} \in X_{(\infty)}$. Finally, let

$$\Omega_{opt} := \Omega(x_{opt}) > \Omega_{min} \quad \text{and} \quad c > \Omega_{min} \,. \tag{4.12}$$

Thus, $\vartheta$ and $\chi$ are given so that both constraints of problem (4.6) are essential. Then, the inequality $\|Ax_{(\infty)}-z\| - \chi\sqrt{\Omega(x_{(\infty)})} > \vartheta$ holds for all $x_{(\infty)} \in X_{(\infty)}$. O

Remark 4.10:
If the first inequality of (4.12) were injured, the observation data vector z would be so hopelessly corrupted as to make any identification at all ill-advised. In such a case, the observation data would fail to provide any new information in addition to the deterministic a priori information expressed by D. O

Theorem 4.11:
Consider a regularized solution $x_{(\alpha)} \in X_{(\alpha)}$, where $\alpha > 0$ is the regularization parameter. If we have

$$\|Ax_{(\alpha)}-z\| = \vartheta + \chi\sqrt{\Omega(x_{(\alpha)})} \quad , \tag{4.13}$$

then $x_{(\alpha)}$ is a solution to problem (4.6), i.e., $x_{(\alpha)}=x_{opt}$. However, if

$$\Omega(x_{(\alpha)}) = c \,, \tag{4.14}$$

then $x_{(\alpha)}$ is a solution to (4.7), i.e., $x_{(\alpha)}=x_c$. O

Proof:
Provided that (4.13) holds, we suppose, for certain $\tilde{x} \in D$, $\Omega(\tilde{x}) < \Omega(x_{(\alpha)})$ and $\|A\tilde{x}-z\| \leq \vartheta + \chi\sqrt{\Omega(\tilde{x})}$. Then, $\|A\tilde{x}-z\| \leq \vartheta + \chi\sqrt{\Omega(\tilde{x})} < \vartheta + \chi\sqrt{\Omega(x_{(\alpha)})} = \|Ax_{(\alpha)}-z\|$ and $F(\tilde{x}, \alpha) < F(x_{(\alpha)}, \alpha)$. This contradicts the definition of a regularized solution. On the other hand, if $\tilde{\tilde{x}} \in D$ satisfies $\Omega(\tilde{\tilde{x}}) \leq c$ and $\|A\tilde{\tilde{x}}-z\| < \|Ax_{(\alpha)}-z\|$, then (4.14) again implies $F(\tilde{\tilde{x}}, \alpha) < F(x_{(\alpha)}, \alpha)$. This proves the theorem. #

Hence, we know that regularized solutions are particular di-optimal and dc-optimal solutions. Unfortunately, a converse of Theorem 4.11 requires strong additional conditions. For general nonlinear problems, it may occur that the family of regularized solutions and the sets $X_{opt}$ or $X_c$ are disjoint. This phenomenon is further investigated within this chapter (see also [212]). Due to formula (4.12), from Lemma 4.8 it follows:

Corollary 4.12:

If one of the conditions $I_{ls} = \emptyset$ and $\chi>0$, $I_{ls} = \emptyset, \chi = 0$ and $\mu<\vartheta$ or

$$I_{ls} \neq \emptyset \quad \text{and} \quad \tau_0 := \mu - \chi\cdot\sqrt{\min_{x\in I_{ls}} \Omega(x)} < \vartheta \tag{4.15}$$

is satisfied, then there exists a couple $0<\alpha_1<\alpha_2<\infty$ of regularization parameters with

$$s_{dif}(\alpha_1)=\|Ax_{(\alpha_1)}-z\|-\chi\cdot\sqrt{\Omega(x_{(\alpha_1)})} \leq \vartheta \leq s_{dif}(\alpha_2)=\|Ax_{(\alpha_2)}-z\|-\chi\cdot\sqrt{\Omega(x_{(\alpha_2)})}.$$

$$\tag{4.16}$$

If either $I_{ls} = \emptyset$ or

$$I_{ls} \neq \emptyset \quad \text{and} \quad c<\min_{x\in I_{ls}} \Omega(x)\ , \tag{4.17}$$

then there exists a couple $0<\alpha_3<\alpha_4<\infty$ of regularization parameters with

$$s_{\Omega}(\alpha_4)=\Omega(x_{(\alpha_4)}) \leq c \leq s_{\Omega}(\alpha_3)=\Omega(x_{(\alpha_3)}) \ . \ \circ \tag{4.18}$$

Theorem 4.13:

If the values $\vartheta, \chi$ and c are given so that the hypotheses of Corollary 4.12 are satisfied, then there is a regularization parameter $\tilde{\alpha}>0$ with $x_{(\tilde{\alpha})} \in I_{opt}$ and a value $\tilde{\tilde{\alpha}}>0$ with $x_{(\tilde{\tilde{\alpha}})} \in I_c$ whenever, for all $\alpha>0$, the functional $F(x,\alpha)$ has a unique minimizer with respect to x in D. $\circ$

Proof:

From the Lemmas 4.7 and 4.8 it follows that $s_{res}(\alpha)$, $s_{\Omega}(\alpha)$ and $s_{dif}(\alpha)$ are continuous functions with respect to $\alpha>0$ whenever $x_{(\alpha)}$ is uniquely determined for all positive regularization parameters. The inequalities (4.16) and (4.18) ensure that positive parameters $\tilde{\alpha}$ and $\tilde{\tilde{\alpha}}$ exist such that the equations (4.13) and (4.14) are satisfied. Then, Theorem 4.11 applies. This completes the proof. #

In Remark 4.2, sufficient conditions for the uniqueness of $x_{(\alpha)}, \alpha>0$, are given. However, there are two pathological cases. Firstly, if $x_{(\alpha)}$ fails to be uniquely determined for all $\alpha>0$ and nonlinear inverse problems are under consideration, it may occur that the hypotheses of Corollary 4.12 are satisfied, but positive regularization parameters $\alpha_{opt}$ and $\alpha_c$ exist with

$$\min_{x\in X_{(\alpha_{opt})}} \left\{\|Ax-z\|-\chi\cdot\sqrt{\Omega(x)}\right\}<\vartheta<\max_{x\in X_{(\alpha_{opt})}} \left\{\|Ax-z\|-\chi\cdot\sqrt{\Omega(x)}\right\}, \quad X_{(\alpha_{opt})}\cap I_{opt} = \emptyset \tag{4.19}$$

or

$$\min_{x\in X_{(\alpha_c)}} \Omega(x) < c < \max_{x\in X_{(\alpha_c)}} \Omega(x) \ , \quad X_{(\alpha_c)} \cap X_c = \emptyset \ . \tag{4.20}$$

Such a phenomenon reminds us of the duality gaps in nonlinear, nonconvex optimization, since nonuniqueness of regularized solutions corresponds to nonconvex functions $F(x,\alpha)$. The study of the next section confirms this conjecture. If (4.19) or (4.20) holds, then $x_{opt}$ or $x_c$ cannot be replaced by a regularized solution, but the optimal solutions may be approximated by using pairs $(\alpha_1;\alpha_2)$ and $(\alpha_3;\alpha_4)$ of regularization parameters satisfying (4.16) and (4.18). Then, the quality of approximations can be given by the formulae

$$\Omega(x_{(\alpha_2)}) \leq \Omega(x_{opt}) \leq \Omega(x_{(\alpha_1)}) \leq \Omega(x_{opt}) + \frac{(\vartheta+\chi\sqrt{\Omega(x_{(\alpha_1)})})^2 - \|Ax_{(\alpha_1)}-z\|^2}{\alpha_1} \tag{4.21}$$

and

$$\|Ax_{(\alpha_3)}-z\|^2 \leq \|Ax_c-z\|^2 \leq \|Ax_{(\alpha_4)}-z\|^2 \leq \|Ax_c-z\|^2 + \alpha_4(c-\Omega(x_{(\alpha_4)})) \ , \tag{4.22}$$

which both result from the definition of regularized solutions. Secondly, $(\bigcup_{\alpha>0} X_{(\alpha)}) \cap X_{opt} = \emptyset$ or $(\bigcup_{\alpha>0} X_{(\alpha)}) \cap X_c = \emptyset$ may also occur whenever the hypotheses of Corollary 4.12 are injured. Then, in the identification case, the formal extension of regularized solutions to nonpositive regularization parameters $\alpha \leq 0$ and, in the control case, the consideration of least squares solutions may help to overcome the theoretic difficulties. However, the arising optimization problems (4.5) and (4.2) with $\alpha < 0$, in general, fail to be well-posed on $R^n$. Therefore, their utility for computing di-optimal and dc-optimal solutions is doubtful. Properties and examples according to regularized solutions with negative regularization parameters are presented in [212, §6].

### 4.1.2. Regularization as a Lagrange Multiplier Method

The structure of Tikhonov's functional $F(x,\alpha)$ suggests a relationship to a Lagrange multiplier methods. We try to formulate the mathematical relations between the Lagrangian saddle-point problems according to (4.6) and (4.7), respectively, and Tikhonov's regularization method. However, by using the quasimonotonicity of $\Omega$ (see Assumption 4.9), let us beforehand present some equivalent formulations to (4.6).

**Lemma 4.14:**

The optimization problems (4.6) and

$$\underset{\substack{x\in D \\ \|Ax-z\| = \vartheta+\chi\sqrt{\Omega(x)}}}{\text{minimize}} \ \Omega(x) \tag{4.23}$$

are equivalent. Moreover, there is a real number $\varrho>0$ such that

$$\begin{aligned} &\underset{x\in D}{\text{minimize}}\ \Omega(x) \\ &\|Ax-z\|^2=\varrho\cdot\Omega(x) \end{aligned} \tag{4.24}$$

is also equivalent to the problems (4.6) and (4.23). o

Proof:

The equivalence of problems (4.6) and (4.23) is a direct consequence of the quasimonotonicity of $\Omega$. Let exist a solution $x_{opt}$ to (4.6) with $\|Ax_{opt}-z\|<\vartheta+\chi\cdot\sqrt{\Omega(x_{opt})}$. Then, in view of the left inequality of (4.12) we have $\Omega(x_{opt})>\Omega_{min}$ and, due to the continuity of A and $\Omega$, there is a sphere $S=\bar{S}(x_{opt},\varepsilon)\subseteq R^n$ with $\|Ax-z\|<\vartheta+\chi\cdot\sqrt{\Omega(x)}$ for all $x\in S\cap D$. However, we know that $x_{opt}$ fails to be a local minimizer of $\Omega$ in D. Consequently, there exists a vector $x^{(1)}\in S\cap D$ with $\Omega(x^{(1)})<\Omega(x_{opt})$. This contradicts the optimality of $x_{opt}$. Hence, (4.6) and (4.23) are equivalent optimization problems. Now we are going to show that $x_{opt}$ as a solution to (4.23) also is a minimizer to (4.24). Set

$$\varrho:=\frac{(\vartheta+\chi\cdot\sqrt{\Omega_{opt}})^2}{\Omega_{opt}}\ , \tag{4.25}$$

where owing to formula (4.12) $\Omega_{opt}:=\Omega(x_{opt})>0$. So, we have

$$\|Ax_{opt}-z\|^2=(\vartheta+\chi\sqrt{\Omega(x_{opt})})^2=(\vartheta+\chi\cdot\sqrt{\Omega(x_{opt})})^2\cdot\frac{\Omega(x_{opt})}{\Omega_{opt}}=\varrho\cdot\Omega(x_{opt}).$$

Hence, $x_{opt}$ belongs to the set of feasible solutions of problem (4.24). Conversely, there is no $x^{(2)}\in D$ with $\|Ax^{(2)}-z\|^2=\varrho\cdot\Omega(x^{(2)})$ and $\Omega(x^{(2)})<\Omega_{opt}$. Namely, $\|Ax^{(2)}-z\|^2=(\vartheta^2+\chi^2\Omega_{opt}+2\vartheta\chi\sqrt{\Omega_{opt}})\cdot$

$$\frac{\Omega(x^{(2)})}{\Omega_{opt}}\leq\vartheta^2+\chi^2\cdot\Omega(x^{(2)})+2\vartheta\chi\sqrt{\frac{\Omega(x^{(2)})}{\Omega_{opt}}}\cdot\sqrt{\Omega(x^{(2)})}\leq(\vartheta+\chi\cdot\sqrt{\Omega(x^{(2)})})^2,$$

violating the optimality of $x_{opt}$ with respect to (4.6). Therefore, $x_{opt}$ is also a solution to problem (4.24). Finally, assume $x^{(3)}\in D$ to solve (4.24). Then the above reflection yields $\Omega(x^{(3)})=\Omega_{opt}$.

Moreover, with $\|Ax^{(3)}-z\|^2=\dfrac{(\vartheta+\chi\sqrt{\Omega_{opt}})^2}{\Omega_{opt}}\cdot\Omega(x^{(3)})=(\vartheta+\chi\cdot\sqrt{\Omega(x^{(3)})})^2$,

the vector $x^{(3)}$ proved to be an optimal solution to problem (4.23). This completes the proof. #

Now we are going to formulate the Lagrangian functionals and saddle-point problems concerning (4.6) and (4.7). For theoretic aspects of nonlinear optimization, in general, compare e.g. GROSSMANN/KLEINMICHEL [183]. The Lagrangian functional according to problem (4.24) is

$$L(x,\lambda)=\Omega(x)+\lambda\cdot(\|Ax-z\|^2-\varrho\cdot\Omega(x))\ ,\ \lambda\geq 0\ . \tag{4.26}$$

We are interested in finding all pairs $(\tilde{x};\tilde{\lambda}) \in D\times[0,\infty)$ satisfying the inequalities

$$L(\tilde{x},\lambda) \leq L(\tilde{x},\tilde{\lambda}) \leq L(x,\tilde{\lambda}) \text{ , for all } x\in D \text{ and } \lambda\geq 0, \tag{4.27}$$

i.e., we ask for all saddle-points of the Lagrangian saddle-point problem associated with (4.24). It is well-known (see [183, p.24]) that any vector $\tilde{x}\in D$ of such a saddle-point problem is a solution to the optimization problem (4.24) and hence a di-optimal solution $\tilde{x}=x_{opt}$. Due to (4.12) there is no saddle-point $(\tilde{x};0)$ satisfying (4.27). If we exclude $\lambda=0$, then the Lagrangian functional (4.26) can be written in the form

$$L(x,\lambda) = \lambda\left[\|Ax-z\|^2 + \left(\frac{1}{\lambda}-\varrho\right)\Omega(x)\right] = \lambda\cdot F\left(x,\frac{1}{\lambda}-\varrho\right), \quad \lambda>0 \text{ .} \tag{4.28}$$

This implies the following theorem:

Theorem 4.15:

An equivalent formulation to the Lagrangian saddle-point problem associated with (4.24) is to find $(\tilde{x};\tilde{\lambda}) \in D\times(0,\infty)$ such that

$$F\left(\tilde{x},\frac{1}{\tilde{\lambda}}-\varrho\right) \leq F\left(x,\frac{1}{\tilde{\lambda}}-\varrho\right) \text{ , for all } x\in D \text{ ,} \tag{4.29}$$

and

$$\|A\tilde{x}-z\|^2 = \varrho\cdot\Omega(\tilde{x}) \text{ .} \tag{4.30}$$

Consequently, a pair $(\tilde{x},\tilde{\lambda}) \in D\times(0,\infty)$ is a saddle-point if and only if $\tilde{x}=x_{(\alpha)}$ is a regularized solution where the regularization parameter

$$\alpha = \frac{1}{\tilde{\lambda}} - \varrho \tag{4.31}$$

satisfies the equation (4.13). o

There are nonconvex optimization problems, for which the associated Lagrangian saddle-point problems fail to have a solution. These so-called duality gaps are responsible for the first pathological case in approximating di-optimal solutions by regularized solutions (see formula (4.19)). On the other hand, saddle-points may occur if $\tilde{\lambda}>0$, i.e., if $\alpha>-\varrho$. Consequently, negative regularization parameters cannot be excluded a priori. This fact refers to the second pathological case, which arises whenever $X_{ls} \neq \emptyset$ and (4.15) does not hold.

The Lagrangian functional according to problem (4.7)

$$L(x,\lambda) = \|Ax-z\|^2 + \lambda(\Omega(x)-c) \text{ , } \lambda\geq 0 \text{ ,} \tag{4.32}$$

is also a modification to $F(x,\alpha)$. If a pair $(\tilde{x},\tilde{\lambda}) \in D\times[0,\infty)$ satisfies (4.27), then $F(\tilde{x},\tilde{\lambda}) \leq F(x,\tilde{\lambda})$ for all $x\in D$ and $(\tilde{\lambda}-\lambda)(\Omega(\tilde{x})-c) \geq 0$ for all $\lambda\geq 0$.

Theorem 4.16:

An equivalent formulation to the Lagrangian saddle-point problem associated with (4.7) is to find $(\tilde{x},\tilde{\lambda}) \in D\times[0,\infty)$ such that

$$F(\tilde{x}, \tilde{\lambda}) \leq F(x, \tilde{\lambda}) \text{ , for all } x \in D, \tag{4.33}$$

and either

$$\Omega(\tilde{x}) = c \quad \text{if } \tilde{\lambda}>0 \tag{4.34}$$

or

$$\Omega(\tilde{x}) \leq c \quad \text{if } \tilde{\lambda}=0 \tag{4.35}$$

are satisfied. Consequently, a pair $(\tilde{x}, \tilde{\lambda}) \in D\times[0, \infty)$ is a saddle-point either if $\tilde{x}=x_{(\tilde{\lambda})}$ is a regularized solution with $\alpha=\tilde{\lambda}>0$ satisfying the equation (4.14) or if $\tilde{x}=x_{ls}$ is a least squares solution that belongs to the level set $W_c^{\Omega}$. O

Like in the identification case, duality gaps causing a situation (4.20) are possible. Moreover, if $X_{ls} \neq \emptyset$ and $\min_{x \in X_{ls}} \Omega(x) \leq c$ (see Corollary 4.12), then there exists a least squares solution $x_{ls} \in X_{ls}$ so that $(x_{ls};0)$ is a saddle-point. Negative regularization parameters are not even of theoretical importance in the control case.

By summarizing the above study one can state that the two different families of optimization problems (4.6) and (4.7) can be replaced or at least approximated (see formulae (4.21) and (4.22)) by representatives of the family of regularization problems (4.2) in a unified manner. Hoeever, from the point of view of computing $x_{opt}$ and $x_c$ it remains the problem how to find, in practice, appropriate regularization parameter values $\alpha$. Provided that Corollary 4.12 is applicable, the following algorithm for the identification case gives an example how to construct approximations $x_{alg}$ to $x_{opt}$ by computing a series of regularized solutions with different regularization parameters.

Algorithm 4.17:

Step 0. Choose an interval $[\alpha_1, \alpha_2] \subseteq (0, \infty)$ satisfying (4.16), two stopping bounds $\varepsilon_1, \varepsilon_2 > 0$, set $i:=2$ and continue.

Step 1. Set $\alpha := \sqrt{\alpha_1 \cdot \alpha_2}$ and $i:=i+1$. Calculate $x_{(\alpha)}$ and $s_{dif}(\alpha)$. If $\vartheta - \varepsilon_1 < s_{dif}(\alpha) \leq \vartheta$, then set $x_{alg} := x_{(\alpha)}$ and stop; otherwise if $s_{dif}(\alpha) \leq \vartheta - \varepsilon_1$, set $\alpha_1 := \alpha$ and continue; if $s_{dif}(\alpha) > \vartheta$, set $\alpha_2 := \alpha$ and continue.

Step 2. If $\frac{\alpha_2}{\alpha_1} \leq 1 + \varepsilon_2$, set $x_{alg} := x_{(\alpha_1)}$ and stop; otherwise return to Step 1. O

If the interval $[\alpha_1, \alpha_2]$ corresponds to a quotient $\frac{\alpha_2}{\alpha_1} = q > 1$, then in the worst case after computing a maximum number of $i = \text{entier}\left(\frac{\ln q}{\ln(1+\varepsilon_2)}\right) + 3$ various regularized solutions the algorithm terminates, since a regularization parameter $\alpha_{opt}$ (see remarks on formula (4.19) has been found approximately. If no duality gap arises and $\varepsilon_2$ is small enough in comparison with $\varepsilon_1$, then a vector $x_{alg}=x_{(\alpha)}$ with $\vartheta-\varepsilon_1 < s_{dif}(\alpha) \leq \vartheta$ had already been found after computing a fewer number of regularized solutions.

### 4.1.3. Alternative Strategies for Choosing the Regularization Parameter

There are several different possible strategies for choosing the regularization parameter $\alpha$ in problem (4.2). The most famous principle is Morozov's discrepancy principle (see MOROZOV [311], [320], cf. also GROETSCH [176]). For our full-discretization model, this principle in its ordinary form consists in approximating optimal solutions $x_{opt}$ in the sense of Theorem 3.39 by an appropriate regularized solution whenever $\Omega$ is quasimonotonous and a sum $\delta+h$ of observation and discretization error bounds is given. Then $\alpha_{discr}>0$ as the regularization parameter according to the discrepancy principle will solve the problem

$$\underset{\substack{\alpha>0 \\ \|Ax_{(\alpha)}-z\| \leq \delta+h}}{\text{minimize}} \left| \|Ax_{(\alpha)}-z\| - (\delta+h) \right| \quad . \tag{4.36}$$

If values $\vartheta$ and $\chi$ are available such that the problem (4.6) can be considered, then the generalized discrepancy principle with the solution $\alpha_{opt}>0$ of

$$\underset{\substack{\alpha>0 \\ \|Ax_{(\alpha)}-z\| \leq \vartheta+\chi\sqrt{\Omega(x_{(\alpha)})}}}{\text{minimize}} \left| \|Ax_{(\alpha)}-z\| - (\vartheta+\chi\cdot\sqrt{\Omega(x_{(\alpha)})}) \right| \tag{4.37}$$

as a characteristic regularization parameter applies in order to approximate di-optimal solutions $x_{opt}$. From studies of the previous sections one can derive the behaviour of regularized solutions, for which $\alpha$ is chosen by the discrepancy principle or its generalization. Whenever duality gaps are excluded, e.g. if A is linear and $\Omega$ convex, the problems (4.36) and (4.37) can be written in the simpler forms

$$\alpha_{discr} : \|Ax_{(\alpha)}-z\| = \delta+h \ , \alpha>0 \quad , \tag{4.38}$$

and

$$\alpha_{opt} : \|Ax_{(\alpha)}-z\| = \vartheta+\chi\cdot\sqrt{\Omega(x_{(\alpha)})} \ , \alpha>0 \quad . \tag{4.39}$$

Let us come back now to the modelling situation of Sec. 3.3.. We are going to identify a fixed vector $x^* \in D$. Since the values $\delta$ and h are given error majorants, the value $\alpha_{discr}$ subject to (4.38) tends to overestimate a desired value $\alpha$ in the following sense. In general, we have for sufficiently pessimistic bounds $\delta$ and h $\|Ax^*-z\| < \|Ax_{(\alpha_{discr})}-z\|$. Owing to the inequality $\|Ax^*-z\|^2 + \alpha_{discr}\Omega(x^*) \geq$

$\geq \|Ax_{(\alpha_{discr})} - z\|^2 + \alpha_{discr} \cdot \Omega(x_{(\alpha_{discr})})$, this implies $\Omega(x^*) > \Omega(x_{(\alpha_{discr})})$. As we know, $s_{\Omega}(\alpha)$ grows if $\alpha$ decreases. Therefore, a value $\alpha < \alpha_{discr}$ would better enable us to approximate $\Omega(x^*)$ by $\Omega(x_{(\alpha)})$. If $\Omega$ is a smoothing functional (see formula (2.95)), e.g. if fewer oscillating vectors x yield smaller values $\Omega(x)$ than vectors that considerably vary in their components, then the discrepancy principle oversmoothes the real solution. That is, $x_{(\alpha_{discr})}$ is too smooth in comparison with $x^*$. Morozov's discrepancy principle can be modified if stochastic noisy data problems are considered. Then, the oversmoothing my be reduced by the introduction of "fudge factors". For the problem of choosing the regularization parameter value from the statistical viewpoint, we refer to Chapter 5 (see formula (5.51)).

If the bounds $\delta$ and h tend to get more pessimistic, the regularized solutions $x_{(\alpha_{discr})}$ change for the worse. In the extreme case, there are no error bounds available. Then, a good way is to choose $\alpha$ interactively, trying several different values and picking that one that yields a best interpretable regularized solution (see ELDÉN [102]). However, there are also some heuristic criteria, which are motivated under special assumptions (see FRIEDRICH et al. [138, p.31], also [198]) and which proved their utility in applications (see [213], [436-37] and [486]). These parameter choice criteria do not require any error bound information.

Thus, we have the quasioptimality criterion that selects $\alpha_{qo}$ as a solution to

$$\underset{\alpha>0}{\text{minimize}} \left\| \alpha \cdot \frac{dx}{d\alpha}(\alpha) - \right\| \tag{4.40}$$

(see TIKHONOV/ARSENIN [434, Chap.II, §7] and BAGLAI [19]). Really, for a wide class of discretized identification problems, the functions $\|\alpha \frac{dx}{d\alpha}(\alpha) - \|$ and $\|x^* - x_{(\alpha)}\|$ are decreasing with $\alpha$ if $\alpha$ is small and increasing with $\alpha$ if $\alpha$ is large. In addition, the arguments $\alpha$ associated with the minima of both functions frequently are almost equal. To avoid the determination of derivatives, the criterion (4.40) is often replaced by its discrete version

$$\underset{\substack{\alpha_i = \alpha_0 \cdot 2^{-i} \\ i > 0,\ \alpha_0 > 0}}{\text{minimize}} \|x_{(\alpha_{i+1})} - x_{(\alpha_i)}\| , \tag{4.41}$$

(see TIKHONOV/MOROZOV [438]).

Some authors also suggest choosing $\alpha$ by a quotient criterion

$$\underset{\alpha>0}{\text{maximize}} \quad \frac{\| A(\alpha \cdot \frac{dx}{d\alpha}(\alpha)) - (Ax_{(\alpha)} - z) \|}{\| Ax_{(\alpha)} - z \|} \tag{4.42}$$

(see e.g. [138, p.32]).

All these heuristic strategies have been used successfully for regularizing inverse problems. However, under weak assumptions, the usefulness of these criteria (4.40) through (4.42) is not guaranteed. The practical realization of any parameter choice strategy requires a search algorithm (see e.g. Algorithm 4.17) based on a series of computed regularized solutions.

At the end of this section let us note that appropriate regularization parameters have to be rather small, in general. This is due to the fact that the regularization problem should be a neighbouring problem to the discretized inverse problem (2.4) or (2.5). On the other hand, the smaller $\alpha$ becomes, the more sensitive the regularized solution is to changes in the input data. Therefore, an appropriate regularization parameter yielding good regularized solutions always arises from a satisfactory trade-off between approximation and stability.

## 4.2. Multi-Parameter Regularization

In this section, consider the discretized identification problem (2.4) as a problem of measurement data interpretation under somewhat extended assumptions. The interpretation of indirect measurements leads to satisfactory tesults if the observation error behaviour receives enough attention. Up to Sec. 4.1., postulated by Assumption 2.64, we have supposed that we know only a value $\delta$ majorizing the Euclidean norm of the measurement error vector y (see formula (2.97)). However, there are identification problems, for which the data vector z consists of very heterogeneous components or subvectors. For example, if the inverse problem is expressed mathematically by a coupled system of equations (see e.g. the case of integral equations in Sec. 2.2.6.), then the right-hand side of any equation contributes to the data vector. All these various right-hand sides can represent very different physical objects and attain values of very different order of magnitude. In such a case, it is not recommended to express the measurement precision by one scalar value only. This suggests choosing the following model:

Assumption 4.18:

Throughout this Section 4.2, assume the operator $A: D \subseteq R^n \to R^m$ to be composed of k continuous operators $A_i: D \subseteq R^n \to R^{m_i}$, $i=1(1)k$, $\sum_{i=1}^{k} m_i = m$ such that $Ax = \begin{pmatrix} A_1x \\ A_2x \\ \vdots \\ A_kx \end{pmatrix}$. The data vector $z = \begin{pmatrix} z^{(1)} \\ z^{(2)} \\ \vdots \\ z^{(k)} \end{pmatrix}$ with $z^{(i)} \in R^{m_i}$ and the observation error vector $y = \begin{pmatrix} y^{(1)} \\ y^{(2)} \\ \vdots \\ y^{(k)} \end{pmatrix}$ with $y^{(i)} \in R^{m_i}$, $i=1(1)k$, are subdivided analogously.

Now let be known k independent positive observation error bounds $\delta_1, \delta_2, \ldots, \delta_k$ such that

$$\|y^{(i)}\| \leq \delta_i \, , \quad i=1(1)k \, . \tag{4.43}$$

For the study of this section, the discretization errors are not considered particularly. o

Remark 4.19:

The mathematical model postulated by Assumption 4.18 also involves the interesting special case of m independent error bounds. In such a case, for each component of z a particular error bound is to be prescribed. Then, the Euclidean norm in (4.43) is simplified to the absolute value as follows:

$$|y_i| \leq \delta_i \, , \quad i=1(1)m \, . \tag{4.44}$$

For the multi-bound model of this section, the closed set

$$Z_{\delta,mult} := \{ x \in D: \|A_i x - z^{(i)}\|^2 \leq \delta_i^2 \quad \text{for all } i=1(1)k \} \tag{4.45}$$

describes the set of data-compatible vectors. Throughout this section, let this set be nonempty. o

Definition 4.20:

Any global solution $x_{opt,mult} \in D$ to the optimization problem

$$\underset{x \in Z_{\delta,mult}}{\text{minimize}} \; \Omega(x) \tag{4.46}$$

is called optimal solution to the multi-bound identification problem postulated by Assumption 4.18. o

Due to Assumption 3.26 the optimization problem (4.46) can be proven to be well-posed on $R^n$. However, the numerical computation of vectors $x_{opt,mult}$ according to (4.46) is, in general, very difficult. Only in particular cases, direct methods for minimizing $\Omega(x)$ subject to the set of feasible vectors $Z_{\delta,mult}$ are available. For instance, if A is linear, $\Omega$ a quadratic functional and the special situation (4.44) is submitted, then (4.46) becomes a quadratic programming problem, for the solution of which advanced numerical methods exist (see e.g. LAWSON/HANSON [269], SCHITTKOWSKI/STOER [382]). In contrast to the case of linear operators A, nonlinear identification problems are always associated with nonconvex sets of feasible vectors $Z_{\delta,mult}$. So, a generalization of Tikhonov's regularization method to the multi-bound situation is useful. Several different error bounds can be taken into consideration by introducing simultaneously several different regularization parameters. We will denote this method by the name multi-parameter regularization.

**Definition 4.21:**

Consider the functional

$$F_{mult}(x,\lambda)=\Omega(x)+\sum_{i=1}^{k}\lambda_i\cdot\|A_ix-z^{(i)}\|^2,\quad \lambda_i\geq 0,\quad i=1(1)k\ . \tag{4.47}$$

Any global solution $x_{[\lambda]}$ of the optimization problem

$$\underset{x\in D}{\text{minimize}}\ F_{mult}(x,\lambda)\ ,\quad \lambda\in R_+^k\ , \tag{4.48}$$

is called multi-parameter regularized solution to the parameter vector $\lambda\in R_+^k$. The lower right-hand corner subscript $\lambda$ of $x$ is written enclosed by brackets [.] instead of parentheses (.) in order to distinguish these solutions from ordinary regularized solutions. ○

Obviously, the optimization problem (4.48) is well-posed on $R^n$ for all $\lambda\in R_+^k$ and possesses a convex set D of feasible vectors. The Lagrangian functional $L(x,\lambda)$ according to problem (4.46) is as follows:

$$L(x,\lambda)=\Omega(x)+\sum_{i=1}^{k}\lambda_i(\|A_ix-z^{(i)}\|^2-\delta_i^2)=F_{mult}(x,\lambda)-\sum_{i=1}^{k}\lambda_i\cdot\delta_i^2. \tag{4.49}$$

**Theorem 4.22:**

A pair $(\tilde{x},\tilde{\lambda})\in D\times R_+^k$ is a saddle-point of the Lagrangian saddle-point problem according to (4.46) if and only if $\tilde{x}=x_{[\tilde{\lambda}]}$ is a multi-parameter regularized solution, i.e., a minimizer of (4.47) in D, satisfying simultaneously the equations

$$\tilde{\lambda}_i(\|A_i\tilde{x}-z^{(i)}\|^2-\delta_i^2)=0\ ,\quad \text{for all } i=1(1)k, \tag{4.50}$$

and

$$\|A_i\tilde{x}-z^{(i)}\|^2\leq\delta_i^2\quad \text{if } \tilde{\lambda}_i=0,\ 1\leq i\leq k\ . \tag{4.51}$$

Any vector $\tilde{x}=x_{[\tilde{\lambda}]}$ fulfilling (4.50) through (4.51) is a solution $x_{opt,mult}$ to problem (4.46). ○

**Proof:**

From the right inequality of (4.27), i.e., $L(\tilde{x},\tilde{\lambda})\leq L(x,\tilde{\lambda})$ for all $x\in D$, it follows together with equation (4.49) that just multi-parameter regularized solutions form the vectors $\tilde{x}$ of the sought saddle-points. The left inequality $L(\tilde{x},\lambda) \leq L(\tilde{x},\tilde{\lambda})$, for all $\lambda \in R_+^k$, can be transformed into $\sum_{i=1}^{k} (\tilde{\lambda}_i - \lambda_i)(\|A_i\tilde{x}-z^{(i)}\|^2 - \delta_i^2) \geq 0$ for all $\lambda \in R_+^k$.

This implies, for all $i=1(1)k$, $(\tilde{\lambda}_i - \lambda_i)(\|A_i\tilde{x}-z^{(i)}\|^2 - \delta_i^2) \geq 0$ and hence the pair of conditions (4.50) and (4.51). Furthermore, a vector $\tilde{x}=x_{[\tilde{\lambda}]}$ satisfying (4.50), (4.51) belongs to the set $Z_{\delta,mult}$. The inequality $F_{mult}(x_{[\tilde{\lambda}]},\tilde{\lambda}) \leq F_{mult}(x,\tilde{\lambda})$, which is valid for all $x \in D$, causes $\Omega(x_{[\tilde{\lambda}]}) \leq \Omega(x) + \sum_{i=1}^{k} \tilde{\lambda}_i(\|A_i x - z^{(i)}\|^2 - \delta_i^2) \leq \Omega(x)$ for all $x \in Z_{\delta,mult}$. Thus, $x_{[\tilde{\lambda}]}$ has been verified to be a solution to problem (4.46). #

<u>Remark 4.23:</u>

A multi-parameter regularized solution $x_{[\lambda]}$ minimizing (4.48), which satisfies an equation $\Omega(x_{[\lambda]}) = c$, is also a modification to the dc-optimal solution $x_c$. Namely, by the arguments of the proof of Theorem 4.11 we obtain $x_{[\lambda]}$ as a solution to the problem

$$\underset{\substack{x \in D \\ \Omega(x) \leq c}}{\text{minimize}} \; \sum_{i=1}^{k} \lambda_i \|A_i x - z^{(i)}\|^2 , \tag{4.52}$$

i.e., as a weighted constrained least squares solution. This interpretation is even useful if no error bounds (4.43) are available. In such a case, weights $\lambda_i$, $i=1(1)k$, may help to equilibrate different orders of magnitude in the absolute values of the data components. The consideration of weighted least squares is also motivated from point of view of statistics (see Chapter 5). Then, $\sqrt{\lambda_i}$ denotes the inverse value of standard deviations associated with the observation error components $y^{(i)}$. ○

Now we are going to derive an iterative method for the solution of problem (4.46) by using multi-parameter regularized solutions. We draw the reader's attention to the condition (4.50) or the equivalent fixed point formulation

$$\tilde{\lambda}_i = \tilde{\lambda}_i \cdot \frac{\|A_i \tilde{x} - z^{(i)}\|^2}{\delta_i^2} , \text{ for all } i=1(1)k . \tag{4.53}$$

Then we choose a small positive number $0<\varepsilon \ll 1$ and consider the infinite sequence $\{\lambda^{(j)} = (\lambda_1^{(j)},\ldots,\lambda_k^{(j)})^T\}_{j=1}^{\infty} \subseteq R_+^k$ according to

$$x^{(j)} := x_{[\lambda^{(j)}]}, \quad j=0,1,2,\ldots , \tag{4.54}$$

$$\lambda_i^{(j+1)} := \lambda_i^{(j)} \cdot \max\left(\frac{\|A_i x^{(j)} - z^{(i)}\|^2}{\delta_i^{\,2}}, \varepsilon\right), \quad \begin{matrix} j=0,1,2,\ldots, \\ i=1(1)k \end{matrix} \tag{4.55}$$

with a given initial vector $\lambda^{(0)}$ having positive components $\lambda_i^{(0)}$, $i=1(1)k$. The simultaneous iteration of x and $\lambda$ represents a fixed point iteration succeeding whenever the transformation $\lambda^{(j)} \rightarrow \lambda^{(j+1)}$ by (4.54) and (4.55) satisfies the assumptions of Banach's fixed point theorem (see e.g. COLLATZ [70], also [39, p.76]). However, it seems to be very difficult to check whether the transformation of Lagrange multipliers (4.55) is a contraction mapping. This is a drawback of the considered approach.

<u>Theorem 4.24:</u>
As $j \rightarrow \infty$, let $\lambda^{(j)} \rightarrow \tilde{\lambda} \in R_+^k$ and $x^{(j)} \rightarrow \tilde{x} \in D$. Then, the pair $(\tilde{x}, \tilde{\lambda})$ forms a saddle-point of the Lagrangian saddle-point problem according to (4.46), and $\tilde{x}=x_{[\tilde{\lambda}]}$ is an optimal solution to the considered multi-bound identification problem. ○

<u>Proof:</u>
The pair $(\tilde{x}, \tilde{\lambda}) \in D \times R_+^k$ of limit vectors satisfies the inequality $F_{mult}(\tilde{x}, \tilde{\lambda}) \leq F_{mult}(x, \tilde{\lambda})$, for all $x \in D$, and the equations $\tilde{\lambda}_i = \tilde{\lambda}_i \cdot \max\left(\frac{\|A_i\tilde{x}-z^{(i)}\|^2}{\delta_i^2}, \varepsilon\right)$, for all $1 \leq i \leq k$. Thus, we have, independently for all i, $\tilde{\lambda}_i = 0$ or $\max\left(\frac{\|A_i\tilde{x}-z^{(i)}\|^2}{\delta_i^2}, \varepsilon\right) = 1$. This implies $\tilde{\lambda}_i = 0$ or $\|A_i\tilde{x}-z^{(i)}\| = \delta_i$ for all $i=1(1)k$. Consequently, (4.50) is valid. Whenever, for a certain subscript number i, $\tilde{\lambda}_i = 0$, then there is a monotonically nonincreasing subsequence $\lambda_i^{(j_\nu)} \rightarrow 0$ as $\nu \rightarrow \infty$. In view of the introduced positive value $\varepsilon$ in formula (4.55), the values $\lambda_i^{(j)}$ get strictly positive for all i and j under consideration. This implies
$\|A_i x^{(j_\nu)}-z^{(i)}\|^2 \leq \delta_i^2$, for all $j_\nu$, and hence $\|A_i\tilde{x}-z^{(i)}\|^2 \leq \delta_i^2$ as required by formula (4.51). Then, the assertion of this theorem is a consequence of Theorem 4.22. #

<u>Remark 4.25:</u>
If the formula (4.55) is replaced by the simpler form

$$\lambda_i^{(j+1)} := \lambda_i^{(j)} \cdot \frac{\|A_i x^{(j)}-z^{(i)}\|^2}{\delta_i^2}, \quad j=0,1,2,\ldots, \quad i=1(1)k, \qquad (4.56)$$

then $(x^{(j)}, \lambda^{(j)}) \rightarrow (\tilde{x}, \tilde{\lambda})$ as $j \rightarrow \infty$ implies $\tilde{x}=x_{[\tilde{\lambda}]}$ and (4.50). However, it might occur that $\|A_i x^{(j_0)}-z^{(i)}\|= 0$ for a certain integer $j_o > 0$. Then, $\lambda_i^{(j)} \equiv 0$, for all $j \geq j_o$ and $\|A_i x^{(j)}-z^{(i)}\|^2 \leq \delta_i^2$ can be ensured for no superscript $j > j_o$. On the other hand, the assertion of Theorem 4.24 remains true if we substitute

$$\lambda_i^{(j+1)} := \lambda_i^{(j)} \cdot \max\left(\frac{\|A_i x^{(j)} - z^{(i)}\|^{\Psi}}{\delta_i^{\Psi}}, \varepsilon\right), \quad \Psi > 0 \text{ aribtrary}, \qquad (4.57)$$

for formula (4.55). Namely, (4.50) and $\tilde{\lambda}_i \cdot (\|A_i\tilde{x} - z^{(i)}\|^{\Psi} - \delta_i^{\Psi}) = 0$, $\Psi > 0$ are equivalent. o

The following algorithm gives an instruction how to approximate a solution $x_{opt,mult}$ to problem (4.46) by the vector $x_{alg}$ representing a special multi-parameter regularized solution.

Algorithm 4.26:

Step 0. Choose an exponent $\Psi > 0$, small positive numbers $\varepsilon$ and $\varepsilon_1$, a starting vector $\lambda^{(0)}$ with all components being positive, and a maximum number $j_{max}$ of iterates. Set $j := 0$ and continue.

Step 1. Calculate $x^{(j)} := x_{[\lambda^{(j)}]}$. If $j = j_{max}$, set $x_{alg} := x^{(j)}$ and stop; otherwise calculate $\lambda^{(j+1)}$ according to formula (4.57) and continue.

Step 2. If $\|\lambda^{(j+1)} - \lambda^{(j)}\| \le \varepsilon_1$, set $x_{alg} := x^{(j)}$ and stop; otherwise set $j := j+1$ and return to Step 1. o

In optimization theory, the interest in processes similar to (4.54) and (4.55) has grown considerably in recent years, where formulae like (4.55) or (4.57) are called Lagrange multiplier estimates. Moreover, refinements of such iterations using penalty terms got the name augmented Lagrangian methods (see GILL et al. [149]). On the other hand, ZAVIALOV et al. (see [487, p.149]) proposed a strategy similar to the above introduced method together with some heuristically motivated modifications for the model situation (4.44). From point of view of regularization TAUTENHAHN [421, p.28] (see also [138, p.74]) considered the simultaneous iteration of $x_{(\alpha)}$ and $\alpha$ in the single-parameter regularization case. Then, our method represents a special secant method, and the exponent $\Psi > 0$ (see (4.57)) determining the rate of convergence according to $\lambda^{(j)} \to \tilde{\lambda}$ as $j \to \infty$ can be chosen optimally in some situations. For alternative methods to handle multi-bound models, compare also GOLDMAN [155].

Numerical experiments concerning the problem of Sec. 2.2.6. (see [190]) with two independent error bounds have indicated that the iteration (4.54), (4.55) tends to converge if $1 \le \Psi \le 3$. However, the sensitivity of the iteration process with respect to an imprecisely computed vector $x^{(j)}$ (incomplete minimization of $F_{mult}(x, \lambda^{(j)})$) considerably grows with $\Psi$.

## 4.3. Descriptive, Adaptive and Iterative Regularization

In this section, we are going to review three strategies which are also reckoned among the regularization methods. All these methods generate solutions that continuously depend on the data. However, they are well-distinguished from Tikhonov's regularization method (see Sec. 4.1.), since they avoid using subjective a priori information based on minimizing an evaluation functional $\Omega$ over a set of data-compatible solutions.

The method of descriptive regularization, in terms of this text, is based on least squares solutions $x_{ls}$ (see problem (4.5)). Contrary to general least squares problems, here the objective a priori information has to prevent or to weaken the instability of the optimization problem, e.g. by forming a compact set D or by lessening the values of the modulus function $\omega_A(-1)(\ .\ ,\ D\ )$ (see (3.61)). This requires using a sufficiently complete description of all qualitative solution properties (lower and upper bounds, piecewise monotonicity and convexity as formulated by (2.66) and (2.67) and more complex vector component relations arising from the physical or technical background). If monotonicity and convexity requirements are used, then Theorem 2.58 justifies the application of descriptive regularization even for large n. As for the literature the reader can be recommended inter alia the papers [99], [155], [156], [158] and the book [441].

The main idea of adaptive regularization is to improve the form of the stabilizing functional by means of already computed regularized solutions. Consider the special Tikhonov regularization defined by the hand of the optimization problem

$$\underset{x \in D}{\text{minimize}}\ F_{\bar{x}}(x,\alpha)\ ,\quad \alpha>0\ , \tag{4.58}$$

with

$$F_{\bar{x}}(x,\alpha) = \|Ax-z\|^2 + \alpha\cdot\|x-\bar{x}\|^2\ , \tag{4.59}$$

where $\bar{x}\in R^n$ is a given vector. Then, regularization can be interpreted as a method for finding the, in general, small actual changes of a known mean vector $\bar{x}$ in dependence of the actual data vector $z$ (see also Chapter 5). Adaptive reasoning would suggest recomputing regularized solutions, where in each case the new mean vector equals the regularized solution just obtained. That is, we have a sequence $\{x^{(j)}\}_{j=1}^{\infty}\subseteq D$ of solutions to

$$\underset{x \in D}{\text{minimize}}\ F_{x^{(j-1)}}(x,\alpha)\ ,\quad \alpha>0\ ,\quad j=1,2,\ldots\ , \tag{4.60}$$

respectively, with an initial vector $x^{(0)}\in D$.

The second iterate $x^{(2)}$ may have better properties than $x^{(1)}$ sometimes (see [311], [438]). However, no additional information seems to arise if we continue to compute iterates according to (4.60). The meaning of iterates $x^{(3)}, x^{(4)}, \ldots$ is difficult to interprete.

Theorem 4.27:
A sequence $\{x^{(j)}\}_{j=0}^{\infty}$ of solutions to (4.60) is associated with a convergent nonincreasing residual norm sequence, i.e.,

$$\mu \leq \lim_{i \to \infty} \|Ax^{(i)}-z\| \leq \|Ax^{(j+1)}-z\| \leq \|Ax^{(j)}-z\| \;, \; j=0,1,2,\ldots \;. \quad (4.61)$$

Proof:
We have $\|Ax^{(j+1)}-z\|^2 \leq \|Ax^{(j+1)}-z\|^2+\alpha\cdot\|x^{(j+1)}-x^{(j)}\|^2 \leq \|Ax^{(j)}-z\|^2+\alpha\cdot\|x^{(j)}-x^{(j)}\|^2 = \|Ax^{(j)}-z\|^2$. Hence, the nonnegative residual norms under consideration do not increase. Therefore, $\{\|Ax^{(j)}-z\|\}_{j=0}^{\infty}$ is a convergent sequence of real numbers with $\alpha\cdot\|x^{(j+1)}-x^{(j)}\|^2 \to 0$ as $j \to \infty$. Consequently, the changes $\|x^{(j+1)}-x^{(j)}\|$ tend to zero as $j \to \infty$. However, $\{x^{(j)}\}_{j=0}^{\infty}$ is not necessarily a convergent sequence. #

The adaptive regularization process (4.60) is aimed at the possibly incomplete minimization of $\|Ax-z\|^2$ over D. With growing superscript number j, the value $\alpha\cdot\|x^{(j+1)}-x^{(j)}\|^2$ and therefore the influence of the stabilization term decays.

Finally, the iterative regularization method completely gives up the use of a stabilizing functional. In this method, attention is paid to an iteration process

$$x^{(j+1)} := x^{(j)} - \gamma_j \cdot s^{(j)} \;, \quad j=0,1,2,\ldots, \quad (4.62)$$

aimed at solving (4.5). The steplengthes $\gamma_j > 0$ and direction vectors $s^{(j)} \in R^n$ must be chosen appropriately at any iteration step. Let $x^{(j)} \in D$, for all j, moreover, $\lim_{j\to\infty} \|Ax^{(j)}-z\| = \mu := \inf_{x \in D} \|Ax-z\|$. The sequence $\{\|Ax^{(j)}-z\|\}_{j=0}^{\infty}$ is assumed to be strictly decreasing. In this context, instability is avoided by controlling the number of really performed iteration steps. We suppose that (2.97) holds, furthermore that the discretization error is negligible and finally that $\|Ax^{(0)}-z\| > \delta$. Then as a regularized solution we pick the one iterate $x^{(\tilde{j})}$ satisfying

$$\|Ax^{(\tilde{j})}-z\| \leq \delta < \|Ax^{(\tilde{j}-1)}-z\| \quad . \quad (4.63)$$

A regularizing effect is achieved by terminating the iteration before the unwanted irregular part of the solution has converged. For linear unconstrained inverse problems, the relations of iterative regularization may be expressed by convincing arguments in terms of singular values (see Sec. 4.4.).

The stopping criterion (4.63) is an analogue of Morozov's discrepancy principle for Tikhonov's regularization method. We obtain approximate solutions lying in the set $Z_\delta$, which are compatible with the data. However, in contrast to the case of Tikhonov regularization, in iterative regularization no subjective a priori information is exploited for the selection of a special solution vector of $Z_\delta$. The solution $x^{(\tilde{j})}$ obtained depends on the choice of $\gamma_j$ and $s^{(j)}$, for $j=0,1,2,\ldots,\tilde{j}-1$. Therefore, the properties that distinguish the vector $x^{(\tilde{j})}$ from all other vectors of $Z_\delta$ are somewhat random. In this sense, iterative regularization is an uncontrolled form of regularization. But the amount of computation of regularized solutions according to (4.63) tends to be small in comparison with the expense for realizing the discrepancy principle in Tikhonov regularization via a search algorithm (see Algorithm 4.17). For more detailed studies on iterative regularization (also termed iterative filtering), we refer e.g. to LANDWEBER [259], BJÖRCK/ELDÉN [33] and BAKUSHINSKI [21].

## 4.4. Regularization of Unconstrained Problems

### 4.4.1. Euler's Equation

All propositions and notes stated within the previous sections of this chapter remain true if we consider the regularization of unconstrained discretized inverse problems. However, some difficulties disappear since the associated optimization problem (4.2) and as $\alpha$ tends to zero (4.5) are unconstrained minimization problems. Moreover, the linear case studied in Sec. 4.4.2. will be used predominantly in order to make quite clear the mode of operation of the regularization method.

Assumption 4.28:

Throughout this Section 4.4., assume $D = R^n$ and the operator $A: R^n \to R^m$ to be continuously Fréchet-differentiable over $R^n$, where

$$A'(x) = \left[\frac{\partial (Ax)_i}{\partial x_j}\right]_{\substack{i=1(1)m \\ j=1(1)n}}$$ denotes the Fréchet derivative at a

point x. Moreover, suppose the variety of evaluation functionals $\Omega(x)$ to be confined to quadratic forms

$$\Omega(x) = (x-\bar{x})^T \Phi (x-\bar{x}) = (\Phi(x-\bar{x}), x-\bar{x}) \qquad (4.64)$$

with $\bar{x} \in R^n$ a given vector and $\Phi \in R^{n \times n}$ a given symmetric positive definite matrix satisfying the inequality

$$(\Phi x, x) \geq \gamma \cdot \|x\|^2 \text{ , for fixed } \gamma > 0 \text{ and all } x \in R^n. \quad \circ \qquad (4.65)$$

Remark 4.29:

Almost all frequently used subjective a priori information may be expressed in a quadratic form (4.64) for appropriately chosen $\bar{x}$ and $\Phi$ . If smooth functionals (2.94) are discretized as (2.95) indicates, the associated matrix $\Phi$ is tridiagonal (for details, see e.g. [138, p.23]). If second derivative $L_2$-norm squares are added in $\Pi$ , the matrix $\Phi$ becomes fivediagonal. For third derivatives, sevendiagonal matrices are obtained and so on. As for a stochastically motivated nonquadratic evaluation functional of some importance, we refer to [138, p.75] and Sec.5.1.4. (formula (5.77)). ○

Lemma 4.30:

Due to Assumption 4.28 we have

$$\nabla \|Ax\|^2 = 2\cdot(A'(x))^T\cdot Ax \text{ , for all } x\in R^n \text{ . } \circ \qquad (4.66)$$

Proof:

In view of $\nabla\|x\|^2 = 2x$ , the relation (4.66) is a direct consequence of the chain rule in differential calculus (see e.g. [394, p.30]).#

Theorem 4.31:

Under the conditions stated above in Assumption 4.28 let $x=x_{(\alpha)}$ be a regularized solution minimizing (4.1) over $R^n$. Then, this vector satisfies Euler's equation

$$(A'(x))^T(Ax-z) + \alpha\cdot\Phi(x-\bar{x}) = 0 \text{ .} \qquad (4.67)$$

On the other hand, if $x=x_{ls}$ is a least squares solution minimizing $\|Ax-z\|^2$ over $R^n$, then

$$(A'(x))^T(Ax-z) = 0 \text{ . } \circ \qquad (4.68)$$

Proof:

The necessary first order condition for a vector x to be a minimiser of $F(x,\alpha)$ is $\nabla_x F(x,\alpha) = 0$ . If we consider the Cholesky factorization

$$\Phi = L\cdot L^T \text{ , } L\in R^{n\times n} \text{ lower triangular ,} \qquad (4.69)$$

then $F(x,\alpha) = \|Ax-z\|^2 + \alpha\cdot\|L^T(x-\bar{x})\|^2$ . Since a linear operator and its Fréchet derivative coincide when matrix representations are used, (4.67) and (4.68) result from Lemma 4.30. #

The vector equation (4.67) for regularized solutions and its limiting case (4.68) describe systems of n , in general nonlinear, equations in n unknowns. Note that due to the well-posedness of problem (4.2) the system (4.67) has at least one solution as contrasted with the system (4.68), which may be inconsistent whenever no least squares solutions exist. On the other hand, (4.67) is uniquely solvable if the residual norm square is convex (cf. condition (4.4)). Whenever (4.67) possesses a unique solution vector x, then this vector is the uniquely determined minimizer $x_{(\alpha)}$ of $F(x,\alpha)$.

There are iterative methods for the solution of Euler's equation (4.67) (see TAUTENHAHN [421]). Consider an iteration process, where the (j+1)-th iterate $x^{(j+1)}$ minimizes the functional

$$F^{(j+1)}(x,\alpha):=\|Ax^{(j)}+A'(x^{(j)})(x-x^{(j)})-z\|^2+\alpha(\Phi(x-\bar{x}),x-\bar{x}). \quad (4.70)$$

This functional arises from $F(x,\alpha)$ by a linearization of Ax at a point $\tilde{x}\in R^n$

$$Ax = A\tilde{x} + A'(\tilde{x})(x-\tilde{x}) \quad (4.71)$$

such that $\tilde{x}=x^{(j)}$ equals the j-th iterate.

<u>Lemma 4.32:</u>
A vector x is solution to the unconstrained least squares problem

$$\underset{x\in R^n}{\text{minimize}} \{(Vx,x)-2(v,x)\} \quad (4.72)$$

with $v\in R^n$ and $V\in R^{n\times n}$ a symmetric positive definite matrix iff this vector satisfies the uniquely solvable linear algebraic system

$$V\,x = v\,, \quad (4.73)$$

i.e., if

$$x = V^{-1}\,v\,. \quad (4.74)$$

○

Due to this well-known lemma, from the structure of (4.70) it follows by elementary transformations

$$x^{(j+1)}=x^{(j)}-((A'(x^{(j)})^T\cdot A'(x^{(j)})+\alpha\Phi)^{-1}[(A'(x^{(j)})^T(Ax^{(j)}-z)+ + \alpha\Phi(x^{(j)}-\bar{x})]\,. \quad (4.75)$$

For given initial vector $x^{(0)}$, the sequence $\{x^{(j)}\}_{j=0}^{\infty}$ according to the functional (4.70) is uniquely determined.

<u>Theorem 4.33:</u>
If the sequence $\{x^{(j)}\}_{j=0}^{\infty}$ tends to a limit vector $\tilde{x}\in R^n$, then this vector satisfies Euler's equation (4.67) for Tikhonov's regularization method. ○

<u>Proof:</u>
From (4.75) it follows
$((A'(\tilde{x}))^T(A'(\tilde{x}))+\alpha\Phi)^{-1}[(A'(\tilde{x}))^T(A\tilde{x}-z)+\alpha\Phi(\tilde{x}-\bar{x})] = 0$ whenever $x^{(j)}\to\tilde{x}$ as $j\to\infty$. This implies $(A'(\tilde{x}))^T(A\tilde{x}-z)+\alpha\Phi(\tilde{x}-\bar{x}) = 0$ . #

Having followed the ideas of POLYAK [354] (see also [421, Sec.2.3.]), the sequence $\{x^{(j)}\}_{j=1}^{\infty}$ has been shown to be locally linearly convergent under additional assumptions about the Fréchet derivative $A'(x)$.

## 4.4.2. Regularization of a System of Linear Algebraic Equations

The regularization of linear unconstrained inverse problems is considered comprehensively by so many authors as to make a survey almost impossible (see e.g. [33], [138], [225] or [457]). Here, we only give a brief refelction on this topic with respect to the discretized problem version.

**Assumption 4.34:**

Let the Assumption 4.28 hold. In addition, suppose A to be linear, i.e., the discretized identification problem (2.4) is written in the form

$$A\,x = b\ ,\ x \in R^n,\ b \in R^m \tag{4.76}$$

as a system of m linear algebraic equations in n unknowns. On the other hand, the discretized control problem (2.5) has the configuration

$$\underset{x\in R^n}{\text{minimize}}\ \|Ax-z\|^2 \tag{4.77}$$

with the quadratic convex functional $\|Ax-z\|^2=(A^TAx,x)-2(x,z)+\|z\|^2$. Furthermore, let

$$A = V\cdot\left[\begin{array}{ccc|c} \sigma_1 & & 0 & \\ & \sigma_2 & & 0 \\ 0 & & \sigma_r & \\ \hline & 0 & & 0 \end{array}\right]\cdot U^T \tag{4.78}$$

be a singular value decomposition of A according to formula (2.8) with $r=\text{rank}(A)$, $U=(u^{(1)}|u^{(2)}|\dots|u^{(n)})\in R^{n\times n}$ forming an orthogonal matrix, $V=(v^{(1)}|v^{(2)}|\dots|v^{(m)})\in R^{m\times m}$ forming an orthogonal matrix and $\left[\begin{array}{ccc|c} \sigma_1 & & 0 & \\ & \sigma_2 & & 0 \\ 0 & & \sigma_r & \\ \hline & 0 & & 0 \end{array}\right]\in R^{m\times n}$ being a block-matrix possessing four blocks (three null-matrix blocks and one diagonal-matrix block of dimension r with the singular values $\sigma_1\geq\sigma_2\geq\dots\geq\sigma_r>0$ as diagonal entries). The columns of U and V are supposed to form complete orthonormal systems in $R^n$ and $R^m$, respectively. ○

Theorem 4.35:

Under the Assumption 4.34 the optimization problem (4.2) of Tikhonov's regularization method is equivalent to the system of linear algebraic equations

$$(A^TA+\alpha\Phi)\, x = A^Tz + \alpha\Phi\bar{x}\ ,\ \alpha>0\ , \tag{4.79}$$

i.e., for all $\alpha>0$,

$$x_{(\alpha)}=(A^TA+\alpha\Phi)^{-1}(A^Tz+\alpha\Phi\bar{x})\equiv\bar{x}+(A^TA+\alpha\Phi)^{-1}A^T(z-A\bar{x}) \tag{4.80}$$

is the uniquely determined regularized solution. o

Proof:

In view of $A'(x)\equiv A$, for linear operators A, the equation (4.79) is a special form of Euler's equation (4.67). The regularity of the symmetric positive definite matrix $A^TA+\alpha\Phi$ due to formula (4.65) justifies the formulation (4.80). #

Remark 4.36:

Owing to the identity

$$(A^TA+\alpha\Phi)^{-1}A^T \equiv \Phi^{-1}A^T(A\Phi^{-1}A^T+\alpha\cdot I)^{-1}\ , \tag{4.81}$$

we can also use the expression

$$x_{(\alpha)}= \bar{x}+\Phi^{-1}A^T(A\Phi^{-1}A^T+\alpha\cdot I)^{-1}(z-A\bar{x}) \tag{4.82}$$

in order to describe the regularized solution. Elementary matrix operations help verify the identity of all three given kinds of notation for $x_{(\alpha)}$. If $m<n$ and $\Phi^{-1}$ is available, then the variant (4.82) may be advantageous. It requires the inversion of a small m-dimensional matrix $A\Phi^{-1}A^T+\alpha\cdot I$, whereas $A^TA+\alpha\Phi$ belongs to $R^{n\times n}$. As to some numerical refinement in the computation of $x_{(\alpha)}$ using bidiagonal matrices as a form of incomplete singular value decomposition, we refer to VOEVODIN [465] and ELDÉN [100] (see also [138, p.64]). o

By the following lemma we summarize some well-known relations between regularization and properties of the Moore-Penrose inverse (cf. KUHNERT [255]).

Lemma 4.37:

The Moore-Penrose inverse (see formula (2.8)) of A is a limit matrix as follows:

$$A^+= \lim_{\alpha\to 0+0}(A^TA+\alpha\cdot I)^{-1}A^T = \lim_{\alpha\to 0+0} A^T(AA^T+\alpha\cdot I)^{-1} \tag{4.83}$$

If A is of full column-rank, then

$$A^+= (A^TA)^{-1}A^T\ ,\ \text{rank}(A)=n\ . \tag{4.84}$$

If A has full row-rank, then

$$A^{+}= A^{T}(AA^{T})^{-1} \quad , \ \mathrm{rank}(A)=m \ . \tag{4.85}$$

Moreover, we have

$$\| A^{T}(A^{T}A+\alpha \cdot I)^{-1} \| \leq \| A^{+} \| \tag{4.86}$$

and

$$\mu := \inf_{x \in R^{n}} \| Ax-z \| = \| (AA^{+}-I)\cdot z \|, \tag{4.87}$$

i.e., $\mu = 0$ iff either rank(A)=m or $z \in \mathcal{R}(A)$ . Finally, $X_{ls} \neq \emptyset$ with the general representation

$$x_{ls}= A^{+}z + x_{N} \ , \quad x_{N} \in \mathcal{N}(A) \text{ arbitrarily chosen} \ , \tag{4.88}$$

for the least squares solutions. From (4.82) it follows

$$\lim_{\alpha \to 0+0} x_{(\alpha)} = \bar{x} + \Phi^{-1/2}(A\Phi^{-1/2})^{+}(z-A\bar{x}) \in X_{ls}^{\Omega} \ . \tag{4.89}$$

In view of Lemma 4.7, the limit vector $\lim_{\alpha \to 0+0} x_{(\alpha)}$ is the only element of the set $X_{ls}^{\Omega}$ and thus the uniquely determined solution to

$$\begin{array}{l} \text{minimize } (\Phi(x-\bar{x}),x-\bar{x}) \\ x \in R^{n} \\ \| Ax-z \| = \| (AA^{+}-I)z \| \end{array} \quad . \ \circ \tag{4.90}$$

The stability of regularised solutions with respect to perturbations in the data z can be expressed by the inequality

$$\| x_{(\alpha)}(z)-x_{(\alpha)}(\tilde{z}) \| \leq \min\left(\frac{\sigma_1}{\alpha\gamma}\| z-\tilde{z} \|, \frac{1}{\sqrt{\gamma}} \| (A\Phi^{-1/2})^{+} \| \cdot \| z-\tilde{z} \|\right) \ , \tag{4.91}$$

where $x_{(\alpha)}(\tilde{z})$ is obtained by substituting $\tilde{z}$ for z in formula (4.80) or (4.82). Namely, we have $\| x_{(\alpha)}(z)-x_{(\alpha)}(\tilde{z}) \| =$ $\| \Phi^{-1}A^{T}(A\Phi^{-1}A^{T}+\alpha \cdot I)^{-1}(z-\tilde{z}) \| = \| \Phi^{-1/2}((A\Phi^{-1/2})^{T}(A\Phi^{-1/2})+\alpha \cdot I)^{-1}\cdot$ $\cdot (z-\tilde{z}) \|$ . If we take into account the formulae (4.65), (4.86) and $\| (A\Phi^{-1}A^{T}+\alpha \cdot I)^{-1} \| \leq \frac{1}{\alpha}$ , we easily derive (4.91). The sensitivity factor $\frac{\sigma_1}{\alpha\gamma}$ , for given $\| A \|$ and $\gamma$, increases if the regularization parameter $\alpha$ decreases. As already noted at the end of Sec. 4.1.3., we suspect that the stability gets lost if $\alpha$ tends to zero. But here, in the linear case, instability in the sense of discontinuous dependence of the solution on the data can be excluded due to $\| (A\Phi^{-1/2})^{+} \| < \infty$ . If, however, the smallest singular value of the matrix $A\Phi^{-1/2}$ is very small, the system (4.79) becomes very ill-conditioned.

Now we are going to study the behaviour of the functions $s_{res}(\alpha)$, $s_{\Omega}(\alpha)$ and $s_{dif}(\alpha)$ in the linear unconstrained case (see Secs. 4.1.1. and 4.1.3.). Owing to (4.82) we have

$$s_{res}^2(\alpha) := \|Ax_{(\alpha)}-z\|^2 = \|\alpha(A\Phi^{-1}A^T+\alpha\cdot I)^{-1}(z-A\bar{x})\|^2 , \qquad (4.92)$$

moreover,

$$s_{\Omega}(\alpha) := \Omega(x_{(\alpha)}) = \|\Phi^{-1/2}A^T(A\Phi^{-1}A^T+\alpha\cdot I)^{-1}(z-A\bar{x})\|^2 . \qquad (4.93)$$

By differentiation of Eq. (4.79) one obtains

$$\frac{dx_{(\alpha)}}{d\alpha} = (A^TA+\alpha\Phi)^{-1}\Phi(\bar{x}-x_{(\alpha)}) . \qquad (4.94)$$

Theorem 4.38:

Provided that Assumption 4.34 holds and

$$A\bar{x}-z \notin N(A^T) \quad , \qquad (4.95)$$

then $s_{res}(\alpha) = \|Ax_{(\alpha)}-z\|$ is a continuous and strictly monotonically increasing function and $s_{\Omega}(\alpha)=(x_{(\alpha)}-\bar{x})^T\Phi(x_{(\alpha)}-\bar{x})$ is a continuous and strictly monotonically decreasing function with respect to $0<\alpha<\infty$. Moreover, we have

$$\lim_{\alpha\to 0+0} s_{res}(\alpha) = \|(AA^+-I)z\|, \qquad \lim_{\alpha\to 0+0} s_{\Omega}(\alpha) = \|(A\Phi^{-1/2})^+(z-A\bar{x})\|^2 \qquad (4.96)$$

and

$$\lim_{\alpha\to\infty} s_{res}(\alpha) = \|A\bar{x}-z\| , \qquad \lim_{\alpha\to\infty} s_{\Omega}(\alpha) = 0 . \qquad (4.97)$$

If, in addition,

$$\|(AA^+-I)z\| - \chi\cdot\|(A\Phi^{-1/2})^+(z-A\bar{x})\| < \delta < \|A\bar{x}-z\| , \qquad (4.98)$$

then there is a uniquely determined regularization parameter $0<\alpha<\infty$ satisfying Eq. (4.13) (generalized discrepancy principle). On the other hand, if

$$0 < c < \|(A\Phi^{-1/2})^+(z-A\bar{x})\| , \qquad (4.99)$$

then there is a uniquely determined regularization parameter $0<\alpha<\infty$ satisfying Eq. (4.14). ○

Proof:

The limit conditions (4.96) and (4.97) arise from (4.89), (4.92) and (4.93). If the strict monotonicity of $s_{res}(\alpha)$ and $s_{\Omega}(\alpha)$ in the above conjectured form can be proven for $0<\alpha<\infty$, then provided (4.98) and (4.99), the existence of a unique positive regularization parameter satisfying Eq. (4.13) and Eq. (4.14), respectively, is evident from the continuity of the functions $s_{res}(\alpha)$ and $s_{\Omega}(\alpha)$.

The continuity, however, follows from the uniqueness of $x_{(\alpha)}$ and from the limit conditions of Lemma 4.8. Now we are going to show the mentioned form of monotonicity for the functions $s_{res}(\alpha)$ and $s_{\Omega}(\alpha)$. By formula (4.92) we obtain

$$\frac{d\, s^2_{res}(\alpha)}{d\alpha} = 2(Ax_{(\alpha)}-z,\ A\cdot\frac{dx}{d\alpha}(\alpha)) = 2(\alpha(A\Phi^{-1}A^T+\alpha\cdot I)^{-1}(z-A\bar{x}),$$
$$(A\Phi^{-1}A^T+\alpha\cdot I)^{-1}(z-A\bar{x})-\alpha(A\Phi^{-1}A^T+\alpha\cdot I)^{-2}(z-A\bar{x})) =$$
$$= 2\alpha((A\Phi^{-1}A^T+\alpha\cdot I)^{-1}(z-A\bar{x}),\ A\Phi^{-1}A^T(A\Phi^{-1}A^T+\alpha\cdot I)^{-2}(z-A\bar{x})) .$$

Powers of the matrices $A\Phi^{-1}A^T$ and $A\Phi^{-1}A^T+\alpha\cdot I$ are commutable. Therefore, $\frac{d\, s^2_{res}(\alpha)}{d\alpha} = 2\alpha((A\Phi^{-1}A^T+\alpha\cdot I)^{-3}(A\Phi^{-1}A^T)^{1/2}(z-A\bar{x})$ , $(A\Phi^{-1}A^T)^{1/2}(z-A\bar{x}))$ is nonnegative due to the positive definiteness of $(A\Phi^{-1}A^T+\alpha\cdot I)^{-3}$. The null-spaces of $A^T$ and $(A\Phi^{-1}A^T)^{1/2}$ obviously coincide. Hence $\frac{d\, s^2_{res}(\alpha)}{d\alpha} > 0$, for all $0<\alpha<\infty$, whenever (4.95) is valid. The function $s_{res}(\alpha)$ strictly grows with $\alpha$ . On the other hand, by formula (4.93) we obtain

$$\frac{d\, s_{\Omega}(\alpha)}{d\alpha} = 2(\Phi^{-1/2}A^T(A\Phi^{-1}A^T+\alpha\cdot I)^{-1}(z-A\bar{x}),-\Phi^{-1/2}A^T(A\Phi^{-1}A^T+\alpha\cdot I)^{-2}\cdot$$
$$\cdot(z-A\bar{x}))= -2((A\Phi^{-1}A^T+\alpha\cdot I)^{-3}(A\Phi^{-1}A^T)^{1/2}\cdot(z-A\bar{x}),(A\Phi^{-1}A^T)^{1/2}(z-A\bar{x}))$$
$$= -\frac{1}{\alpha}\cdot\frac{d\, s^2_{res}(\alpha)}{d\alpha} \leq 0$$

. Consequently, $s_{\Omega}(\alpha)$ possesses the required properties. #

If one remembers the various different strategies for choosing the regularization parameter mentioned in Sec. 4.1.3., some simplifications get possible in the linear unconstrained case. It is well-known that the ordinary discrepancy principle (4.38) can be realized conveniently by Newton's method (see GORDONOVA/MOROZOV [169], FRIEDRICH et al. [138, p.64]). Consider

$$f(\beta) := \|Ax_{(\frac{1}{\beta})}-z\|^2-(\delta+h)^2 = \|(\beta\cdot A\Phi^{-1}A^T+I)^{-1}(z-A\bar{x})\|^2-(\delta+h)^2 .$$

Then, for all $0<\beta<\infty$ , $f(\beta)$ is under the assumptions of Theorem 4.38 a strictly convex and strictly monotonically decreasing function such that a sequence $\{\beta_j\}_{j=0}^{\infty}$ defined as

$$\beta_{j+1} := \beta_j - \frac{f(\beta_j)}{f'(\beta_j)} \qquad (4.100)$$

converges to the unique solution $\beta_{discr} := \frac{1}{\alpha_{discr}}$ of (4.38) whenever $\beta_0 \leq \beta_{discr}$.

The first derivative $f'(\beta)$ may be calculated easily by means of $f'(\beta) = \frac{-1}{\beta^2}\,\frac{d\, s^2_{res}(\alpha)}{d\alpha}$ , $\alpha = \frac{1}{\beta}$ , $\frac{d\, s^2_{res}(\alpha)}{d\alpha} = 2((Ax_{(\alpha)}-z),A\cdot\frac{dx}{d\alpha}(\alpha))$ and (4.94). Formula (4.94) also helps realize the quasioptimality principle (4.40) numerically (see ZAIKIN/MECHENOV [486]

### 4.4.3. A Singular Value Analysis of Regularization

Recalling the uncertainty study of Sec. 3.3., note that we have an explicit upper bound

$$\omega_A(-1)(s, W_c^{\Omega}) := \sup_{\substack{x^{(1)}, x^{(2)} \in R^n \\ \|Ax^{(1)} - Ax^{(2)}\| \leq s \\ (\Phi(x^{(1)}-\bar{x}), x^{(1)}-\bar{x}) \leq c \\ (\Phi(x^{(2)}-\bar{x}), x^{(2)}-\bar{x}) \leq c}} \|x^{(1)} - x^{(2)}\| \leq \begin{cases} \frac{4c}{\gamma} & \text{if } \operatorname{rank}(A) < n \\ \min(\frac{4c}{\gamma}, \frac{s}{\sigma_n}) & \text{if } \operatorname{rank}(A) = n \end{cases} \qquad (4.101)$$

for the values of the modulus functional in the linear unconstrained case. Namely, one can estimate

$$\omega_A(-1)(s, W_c^{\Omega}) \leq \sup_{\substack{x \in R^n \\ \|Ax\| \leq s \\ (\Phi x, x) \leq 4c}} \|x\| \leq \min\Big(\sup_{\substack{x \in R^n \\ (\Phi x, x) \leq 4c}} \|x\|, \sup_{\substack{x \in R^n \\ \|Ax\| \leq s}} \|x\|\Big) =$$

$$= \min\Big(\sup_{\substack{v \in R^n \\ \|v\|^2 \leq 4c}} (\Phi^{-1} v, v), \sup_{\substack{x \in R^n \\ \|Ax\| \leq s}} \|x\|\Big)$$ causing the inequality (4.101).

The uncertainty limits obtained in Sec. 3.3. can be evaluated computationally by using formula (4.101).

Now we are going to analyse the mechanism of regularization by means of the introduced singular value decomposition (4.78). For simplicity of notation, it is assumed that

$$\Phi = I \quad , \bar{x} = 0 \; . \qquad (4.102)$$

This assumption corresponds to a coordinate transformation in the equation (4.79). If we substituted $\tilde{x} = \Phi^{1/2}(x-\bar{x})$, $\tilde{A} = A\Phi^{-1/2}$ and $\tilde{z} = z - A\bar{x}$, i.e., the position of the origin were shifted and the vector basis were linearly transformed, then $(\tilde{A}^T\tilde{A} + \alpha \cdot I)\tilde{x} = \tilde{A}^T\tilde{z}$ , $\alpha > 0$, would express Euler's equation in the new variables.

Provided (4.102) we can derive the singular value expansion

$$x_{(\alpha)} = \sum_{i=1}^{r} \frac{\sigma_i}{\sigma_i^2 + \alpha} \cdot (z, v^{(i)}) \cdot u^{(i)} \; , \quad r = \operatorname{rank}(A) \; . \qquad (4.103)$$

This expression becomes evident if considering the equations $x_{(\alpha)} = (A^TA + \alpha \cdot I)^{-1} A^T (\sum_{i=1}^{m} (z, v^{(i)}) \cdot v^{(i)}) = \sum_{i=1}^{r} \{(z, v^{(i)})(A^TA + \alpha \cdot I)^{-1} \cdot A^T v^{(i)}\}$, $A^T v^{(i)} = \sigma_i u^{(i)}$ and $(A^TA + \alpha \cdot I)^{-1} u^{(i)} = \frac{1}{\sigma_i^2 + \alpha} u^{(i)}$ if $1 \leq i \leq r$, $A^T v^{(i)} = 0$ if $i > r$.

If a vector $x^*$ is to be identified approximately by regularization from the vector $z=Ax^*+y$ of currupted data, where $y$ denotes the unknown perturbation vector, then in view of $(z,v^{(i)})=\sigma_i(x^*,u^{(i)})+(y,v^{(i)})$, $1\leq i\leq r$,

$$x^*-x_{(\alpha)}=\sum_{i=1}^{r}\frac{\alpha}{\sigma_i^2+\alpha}\cdot(x^*,u^{(i)})u^{(i)}+\sum_{i=r+1}^{n}(x^*,u^{(i)})u^{(i)}+ \\ +\sum_{i=1}^{r}\frac{-\sigma_i}{\sigma_i^2+\alpha}\cdot(y,v^{(i)})u^{(i)}. \qquad (4.104)$$

The error vector $x^*-x_{(\alpha)}$ of Tikhonov's regularization method is composed of three terms. On the one hand, the first and second term characterize the error of regularized solutions if the perturbation $y$ is missing. On the other hand, the third term expresses the influence of $y$. The data vector $z$ completely fails to involve information about the null-space components $(x^*,u^{(i)})$, $i>r$, of the sought vector $x^*$, since any null-space vector $x\in N(A)$ correpsonds to the same image vector $Ax=0$. Therefore, the second term does not depend on $\alpha$. The first term, however, shows that the error increases with $\alpha$. Namely, the damping factor $0<\frac{\alpha}{\sigma_i^2+\alpha}<1$ that influences the component $(x^*,u^{(i)})$, $1\leq i\leq r$, grows from 0 to 1 if $\alpha$ passes through the positive half-axis. This would suggest choosing $\alpha$ very small. Although $\alpha$ is not very small, the factor $\frac{\alpha}{\sigma_i^2+\alpha}$ becomes close to zero if $\sigma_i^2$ is large enough. That is, the vector components associated with the large singular values of A are recovered very well by Tikhonov's regularization, whereas regularization fails to identify parts of the vector $x^*$ corresponding to small singular values $\sigma_i$.

This phenomenon is of great importance, since the third term of formula (4.104) yields a substantial contribution to the total error the more the smaller $\alpha$ is chosen. Therefore, a trade-off value $\alpha$ according to one of the principles mentioned in Sec. 4.1.3. has to be selected. Consequently, there are good chances of identifying $x^*$ sufficiently well whenever the components $(x^*,u^{(i)})$ with small integer i (sometimes named principal components) are the dominant ones in $x^*$. Frequently, the oscillation of vectors $u^{(i)}$ increases with growing i. Thus, smooth vectors are preferred in Tikhonov's regularization method in comparison with vectors possessing strongly oscillating components.

At the end of this chapter we come back to adaptive, iterative and generally alternative strategies of regularization subject to the linear unconstrained case. In adaptive regularization, the sequence $\{x^{(j)}\}_{j=0}^{\infty}$ stated by formulae (4.58) through (4.60) appears in the form

$$x^{(j+1)} := x^{(j)} - (A^TA + \alpha \cdot I)^{-1} A^T (Ax^{(j)} - z) \ , \ j=0,1,2,\ldots \ . \qquad (4.105)$$

In addition to Theorem 4.27 note that $\lim_{j\to\infty} x^{(j)} = \tilde{x}$ implies $A^T(A\tilde{x}-z) = 0$ , i.e., $\tilde{x}$ satisfies (4.79) with $\alpha = 0$ and is hence a least squares solution $x_{ls}$ with $\|A\tilde{x}-z\| = \|(AA^+-I)z\| = \mu$ . However, provided $x^{(0)} = 0$, singular value analysis even allows us to verify the proposition below.

<u>Theorem 4.39:</u>
The sequence $\{x^{(j)}\}_{j=0}^{\infty}$ of adaptive regularization is converging to the minimum norm least squares solution, i.e.,

$$\lim_{j\to\infty} x^{(j)} = A^+z \qquad (4.106)$$

with $A^+z$ being the only minimizer of

$$\begin{array}{l} \text{minimize } \|x\| \\ x \in R^n \\ \|Ax-z\| = \mu \end{array} \qquad (4.107)$$

(cf. formula (4.89)). o

<u>Proof:</u>
We have $(x^{(j+1)},u^{(i)}) = ((I-(A^TA+\alpha\cdot I)^{-1}A^TA)x^{(j)},u^{(i)}) + ((A^TA+\alpha\cdot I)^{-1}A^Tz,v^{(i)}) = \frac{\alpha}{\sigma_i^2+\alpha}(x^{(j)},u^{(i)}) + \frac{\sigma_i}{\sigma_i^2+\alpha}\cdot(z,v^{(i)})$, $1\le i\le r$,

and, in view of $x^{(0)} = 0$, $(x^{(j+1)},u^{(i)}) = 0$, $i>r$. Hence, for $1\le i\le r$, $(1-\frac{\alpha}{\sigma_i^2+\alpha})(\tilde{x},u^{(i)}) = \frac{\sigma_i}{\sigma_i^2+\alpha}\cdot(z,v^{(i)})$ and $(\tilde{x},u^{(i)}) = \frac{1}{\sigma_i}(z,v^{(i)})$. However, $(\tilde{x},u^{(i)}) = 0$ if $i>r$. This implies $\tilde{x} = A^+z$ . #

Now pay attention to iterative regularization and consider the special case of (4.62):

$$x^{(j+1)} := x^{(j)} - \rho\cdot A^T(Ax^{(j)}-z) \ , \ x^{(0)} = 0 \ , \ \rho>0 \ , \qquad (4.108)$$

which is due to LANDWEBER (see [259] and [386, p.78]; for modifications, cf. [413]).

<u>Theorem 4.40:</u>
Landweber's iteration sequence $\{x^{(j)}\}_{j=0}^{\infty}$ according to (4.108) satisfies the condition

$$\lim_{j\to\infty} x^{(j)} = A^+z \qquad (4.109)$$

whenever $0<\rho<\frac{2}{\sigma_1^2} = \frac{2}{\|A^TA\|}$. If, in addition, $z \notin N(A^T)$, then

$$\|Ax^{(j+1)}-z\| < \|Ax^{(j)}-z\| \ , \ j=0,1,2,\ldots \ . \qquad (4.110)$$

Provided that $\mu<\sigma<\|z\|$ , there is a uniquely determined $\tilde{j}>0$ satisfying (4.63). o

Proof:

By mathematical induction we can establish

$$x^{(j)}=\sum_{\nu=0}^{j-1}\varrho(I-\varrho A^TA)^{\nu}A^Tz=\sum_{\nu=0}^{j-1}\sum_{i=1}^{r}\varrho\sigma_i(1-\varrho\sigma_i^2)^{\nu}(z,v^{(i)})u^{(i)}=$$

$=\sum_{i=1}^{r}\frac{1-(1-\varrho\sigma_i^2)^j}{\sigma_i}(z,v^{(i)})u^{(i)}$. This formula proves the limit condition (4.109). Namely, $|1-\varrho\sigma_i^2|<1$ and $\lim_{j\to\infty}(1-\varrho\sigma_i^2)^j=0$ whenever $0<\varrho<\frac{2}{\sigma_i^2}$. Then, $x^{(j)}\longrightarrow\sum_{i=1}^{r}\frac{1}{\sigma_i}\cdot(z,v^{(i)})u^{(i)}=A^+z$ as $j\to\infty$.

Moreover, we have $\|Ax^{(j)}-z\|^2=\sum_{i=1}^{r}(1-\varrho\sigma_i^2)^{2j}(z,v^{(i)})^2$. If at least one of the values $(z,v^{(i)})^2$, $i=1(1)r$, is positive, i.e., $z\notin N(A^T)$, then $\|Ax^{(j)}-z\|$ strictly decreases with growing $j$ and the theorem is proven. #

Rewriting the error vector $x^*-x^{(j)}=\sum_{i=1}^{n}(x^*,u^{(i)})u^{(i)}-\sum_{i=1}^{r}\frac{1-(1-\varrho\sigma_i^2)^j}{\sigma_i}(Ax^*+y,v^{(i)})u^{(i)}$, we obtain

$$x^*-x^{(j)}=\sum_{i=1}^{r}(1-\varrho\sigma_i^2)^j(x^*,u^{(i)})u^{(i)}+\sum_{i=r+1}^{n}(x^*,u^{(i)})u^{(i)}$$
$$-\sum_{i=1}^{r}\frac{1-(1-\varrho\sigma_i^2)^j}{\sigma_i}(y,v^{(i)})u^{(i)}\ . \tag{4.111}$$

This formula can be interpreted analogously to (4.104) as to derive the regularizing character of iterative regularization. Here, the role of regularization parameter carries over to the number $j$ of performed iteration steps. A large $j$ is recommended in order to make the first term and therefore $(1-\varrho\sigma_i^2)^j$ small. Note that, for given $j$, the damping effect is again the stronger the larger $\sigma_i$ becomes, i.e., the smaller $i$ is chosen. On the other hand, large values $j$ cause damping factors $1-(1-\varrho\sigma_i^2)^j\approx 1$ regarding the third term. With growing $j$, the unwanted influence of $y$ also grows.

We complete this Chapter 4 with some remarks on the truncated singular value decomposition method for solving ill-conditioned systems of linear algebraic equations. The method (see HANSON [186] and VARAH [456-57]) assumes to know an appropriate small positive number $\tau$ such that vector components $(x,u^{(i)})$ corresponding to singular values $\sigma_i<\tau$ are unwanted. For given $\tau$, the approximate solution to a discretized linear unconstrained problem proposed by this method gets the form

$$x_{\langle\tau\rangle}:=\sum_{\tau\le\sigma_i}\frac{1}{\sigma_i}\cdot(z,v^{(i)})u^{(i)}\ . \tag{4.112}$$

The components of $A^+z$ that correspond to singular values $\sigma_i \geq \tau$ enter the approximate solution $x_{\langle\tau\rangle}$ in an undamped form, whereas they occur with the damping factor $\frac{\sigma_i^2}{\sigma_i^2+\alpha}$ in Tikhonov regularization. The factor $\frac{\sigma_i^2}{\sigma_i^2+\alpha}$ becomes small if $\sigma_i < \tau$. However, in truncated singular value decomposition, the associated components of $A^+z$ are completely ignored.

A numerical computation of $x_{\langle\tau\rangle}$ needs a singular value decomposition of A, i.e., at least the singular values $\sigma_i \geq \tau$ and associated vectors $u^{(i)}$, $v^{(i)}$ are to be made available. Formula (4.112) shows the regularizing character of this method if one derives the analogy estimate to (4.91):

$$\|x_{\langle\tau\rangle}(z)-x_{\langle\tau\rangle}(\tilde{z})\| \leq \frac{1}{\tau}\cdot\|z-\tilde{z}\| . \tag{4.113}$$

Here, the value $\tau$ takes over the role of regularization parameters $\alpha$. But as already mentioned with regard to iterative regularization, the approximate solutions $x_{\langle\tau\rangle}$ thus obtained have somewhat random properties, since no subjective a priori information applies. Aimed at finding a trade-off between stability ($\tau$ large) and approximation to $A^+z$ ($\tau$ small), the value $\tau$ will preferentially be chosen in an interactive manner.

## 5. Regularization of Stochastic Discretized Inverse Problems

### 5.1. The Non-Bayesian Case

#### 5.1.1. Linear Estimators in the Linear Model

It has been fifteen years since the publications of HOERL and KENNARD [191-92] on ridge regression, the stochastic counterpart of Tikhonov's regularization method. From the numerical point of view the stochastic approach to discretized inverse problems was also of interest in the ensuing years (see PETROV [351] , TURCHIN et al. [448], FRANKLIN [130], FRIEDRICH et al. [138] and FEDOTOV [128]). In this chapter, we are going to present some main ideas and propositions concerning the solution of predominantly linear systems of m equations in n unknowns when the data and in the Bayesian case also the solution vector have stochastic character. The majority of assertions to be made holds independently of the condition number of the system matrix, but for ill-conditioned problems it is of great interest to study the interrelations between regularized solutions in the deterministic model (see Chap.4) and best estimates obtained by stochastic reasoning in view of different risk functions (see also [201]).

Linear models of non-Bayesian kind

$$A\,x + y = z\ , \quad A\in R^{m\times n}\ , \ x\in R^n\ , \ y,z\in R^m \tag{5.1}$$

with x being a fixed but unknown vector and y a realization of a centralized random noise vector $\eta$ possessing the moments

$$E\,y = 0\ , \quad \mathrm{cov}(y) = E\,yy^T = C\ , \ C\in R^{m\times m}\ , \tag{5.2}$$

where C denotes a positive definite symmetric covariance matrix, are well-known and have been studied in mathematical statistics for a long time. As for the definition of expected value, expectation vector and covariance matrix, we refer to [272]. Such a model regards the problem of regression, i.e., the regression coefficients $x_i$, i=1(1)n, gathered into the vector x are to be estimated from the observation data vector z. Here, $z\in R^m$ is a realization of the random vector $\zeta = A\,x + \eta$ with $E\,\zeta = A\,x$ . Frequently, in the classical regression model, there are strategies to find an optimal design such that the regressor matrix A gets orthonormal columns. Then, the system of algebraic equations according to (5.1) with $\mathrm{cond}_1(A)= 1$ is ideally conditioned. If orthogonality of $A^TA$ cannot be achieved, there are at least good chances of generating well-conditioned matrices A by an appropriate regressor design.

However, there is a principal distinction between inverse problems that can be modelled by the formulae (5.1) and (5.2) and classical regression. Linear discretized inverse problems with stochastic data are closely related to the so-called quasilinear regression problem described as follows. Given a set of n basis functions $\varphi^{(i)}(t)$, $i=1(1)n$, of the real argument t, a linear combination $x_1\varphi^{(1)}(t)+x_2\varphi^{(2)}(t)+\ldots+x_n\varphi^{(n)}(t)$ is to be found that approximates an observable real function b(t) as best as possible in a certain sense. In order to do this, we choose m arguments $t_1<t_2<\ldots<t_m$ and gather the randomly perturbed observation data $z_j=b(t_j)+y_j$ into the vector z. The matrix A is therefore formed by the entries $(\varphi^{(i)}(t_j))_{i=1(1)n}^{j=1(1)m}$. To a certain extent the condition of A can be controlled by an appropriate selection of arguments $t_j$, $j=1(1)m$. This corresponds to the problem of choosing an optimal experimental design operator $Q^m$ for the discretization of the right-hand side of Eq. (2.1) concerning an inverse problem (see Chap. 2). On the other hand, the basis functions $\{\varphi^{(i)}(t)\}_{i=1}^{n}$ determine the a priori discretization operator $P^n$ of the unknown element in (2.1).

As we have learned from the Secs. 2.2.1. and 2.2.2. concerning the behaviour of linear Fredholm and Volterra integral equations of the first kind, the associated matrix A tends to be ill-conditioned whenever the number n of basis functions is not extremely small. Consequently, randomly perturbed linear discretized inverse problems are concerned with the model (5.1) and ill-conditioned matrices A. Note that the linear regression model (5.1) applies to inverse problems only if the discretization error either is negligible or $Q^m\mathcal{A}\varkappa - AP^n\varkappa$ possesses stochastic character such that the sum of discretization and observation error satisfies (5.2).

The discretization error may be omitted, for instance, if the following situation is under consideration. Let $P^n$ be defined as (2.33) indicates. Moreover, let be known that the observed function $\mathscr{b}=\langle b(s),\ 0\leq s\leq 1\rangle\in L_2[0,1]$ has a small distance from the n-dimensional subspace span$(\mathscr{e}^{(1)},\ldots,\mathscr{e}^{(n)})$ of $L_2[0,1]$ generated by the basis functions $\mathscr{e}^{(i)}=\langle e^{(i)}(s),\ 0\leq s\leq 1\rangle\in L_2[0,1]$, $i=1(1)n$. In this context, the distance mentioned above is called small whenever its value is dominated by the value $\sqrt{E\,\|y\|^2} = \sqrt{\mathrm{trace}(C)}$ of the expected observation error. On the other hand, (5.1) applies if we use the formulae (2.17)-(2.18) in order to discretize a problem (2.16) and the matrix entries $a_{ji}$ of A are considered not exactly in the sense of formula (2.19) but as a weighted mean of the kernel function value $k(s_j,t)$ over a neighbourhood of $t_i$. For simplicity, the kernel function changes are assumed to have random properties such that (5.2) is justified.

Assumption 5.1:
Supposed the linear model (5.1) subject to (5.2), in this section we are going to study linear estimators for recovering either the unknown vector x or linear functionals $(l,x)=l^Tx$, $l\in R^n$ of x from the noisy data vector z. The vector reconstruction problem will be considered from the point of view of identification as well as from the point of view of control. Linear estimators are described by inhomogeneous linear transformations

$$\hat{x}(z) := V\cdot z + v \ , \ V\in R^{n\times m} \ , \ v\in R^n \ , \qquad (5.3)$$

in the vector reconstruction case and by

$$\widehat{l^Tx}(z) := w^Tz + \varrho \ , \ w\in R^m \ , \varrho\in R \ , \qquad (5.4)$$

in the functional reconstruction case. o

Remark 5.2:
We have to distinguish the vector reconstruction case (5.3) and the functional reconstruction case (5.4) in view of the fact that it is considerably easier to find one unknown value $l^Tx$ based on m data than to recover a full vector x using the same information. In consequence, the best estimators with respect to some risk functions may differ for the considered two situations. o

Remark 5.3:
The ingredients V, v, w and $\varrho$ of the estimators (5.3) and (5.4) are selected so that the random influence involved in the data is filtered off as far as possible during the estimation process. Therefore, estimation in models of type (5.1) is frequently termed filtering. Note that Tikhonov's regularization and related techniques (see Chap. 4) also possess filtering properties as the associated singular value analysis of Sec. 4.4.3. has indicated. o

Now let us evaluate various different estimators by a mean square error risk, for the definition of which quadratic loss functions are employed. From the point of view of vector identification the estimator is to be chosen so that

$$E\ \|\hat{x}(z)-x\|^2 = \int_{R^m} \|\hat{x}(z)-x\|^2\cdot p_\eta(y)\cdot dy \qquad (5.5)$$

attains small values in the vector reconstruction case. Here, $p_\eta(y)$ denotes the density functional of the probability distribution of $\eta$. On the other hand, the mean square error risk

$$E\,(1^T\hat{x}(z)-1^Tx)^2 = \int_{R^m} (1^T\hat{x}(z)-1^Tx)^2 \cdot p_\eta(y)\cdot dy \tag{5.6}$$

should be small in the functional reconstruction case. Finally, the control problem consists in finding a vector x from z such that $A\,\hat{x}(z)$ approximates the vector $b := A\,x$, i.e.,

$$E\,\|A\,\hat{x}(z) - A\,x\|^2 = \int_{R^m} \|A\,\hat{x}(z) - A\,x\|^2 \cdot p_\eta(y)\cdot dy \tag{5.7}$$

is to be minimized.

To express the mean square error risks (5.5) through (5.7) in terms of V, v, w and $\varrho$, for linear estimators, we give some interdependences between linear transformations and expected values below. The following lemma is evident from the definition of the expected value.

Lemma 5.4:
Let $\xi$, $\xi^{(1)}$ and $\xi^{(2)}$ be n-dimensional random vectors, the realizations of which we denote by $x$, $x^{(1)}$ and $x^{(2)}$, respectively. Moreover, let $\eta$ be an m-dimensional random vector with realizations y. If

$$\begin{aligned} &Ex = \bar{x}\ ,\ cov(x) = E(x-\bar{x})(x-\bar{x})^T = B\ ,\\ &E\,x^{(1)} = \bar{x}^{(1)},\ cov(x^{(1)}) = B^{(1)},\\ &E\,x^{(2)} = \bar{x}^{(2)},\ cov(x^{(2)}) = B^{(2)},\\ &Ey = 0\ ,\ cov(y) = C\ , \end{aligned} \tag{5.8}$$

then

$$E(x^{(1)}+x^{(2)}) = \bar{x}^{(1)}+\bar{x}^{(2)}, \tag{5.9}$$

$$cov(x^{(1)}+x^{(2)}) = B^{(1)}+B^{(2)} \text{ if } \xi^{(1)} \text{ and } \xi^{(2)} \text{ are independent,} \tag{5.10}$$

$$E\,\|y\|^2 = trace(C)\ , \tag{5.11}$$

$$E\,A\,x = A\,\bar{x}\ ,\quad cov(A\,x) = A\cdot B\cdot A^T \tag{5.12}$$

and

$$E\,1^Tx = 1^T\bar{x}\ ,\ E(1^Tx-1^T\bar{x})^2 = 1^T\cdot B\cdot 1 = (B\cdot 1,\ 1)\ .\ \circ \tag{5.13}$$

Frequently, the mean square errors are expressed my matrix traces. The symbol trace(M), for a quadratic matrix M, designates the sum of the diagonal entries. Before verifying the formulae (5.5)-(5.7) subject to (5.3)-(5.4), we still present the rules for handling matrix traces.

Lemma 5.5:
If $A\in R^{m\times n}$, $V\in R^{n\times m}$ and M is a quadratic matrix of arbitrary dimension, then

$$trace(A\cdot V) = trace(V\cdot A)\ , \tag{5.14}$$

$$\mathrm{trace}(M) = \mathrm{trace}\,(M^T) \tag{5.15}$$

and

$$\mathrm{trace}(A^TA) = \sum_{i=1}^{r} \sigma_i^2 \text{, where } r=\mathrm{rank}(A) \tag{5.16}$$

and the symbols $\sigma_i$ refer to the singular values of A. o

Lemma 5.6:

Provided that linear estimators (5.3) and (5.4) are under consideration, we have

$$E\|\hat{x}(z)-x\|^2 = \|(VA-I)x+v\|^2 + \mathrm{trace}(VCV^T)\ , \tag{5.17}$$

$$E(\widehat{1^Tx}(z)-1^Tx)^2 = ((A^Tw-1,x)+\varrho\,)^2 + (Cw,w) \tag{5.18}$$

and

$$E\|A\hat{x}(z)-Ax\|^2 = \|A((VA-I)x+v)\|^2 + \mathrm{trace}(AVCV^TA^T)\ .\ \text{o} \tag{5.19}$$

Proof:

The difference vectors $\hat{x}(z)-x$ and $A\hat{x}(z)-Ax$ and the difference value $\widehat{1^Tx}(z)-1^Tx$ are of the forms

$\hat{x}(z)-x=VAx+Vy+v-x=((VA-I)x+v)+Vy$,

$A\hat{x}(z)-Ax=AVAx+AVy+Av-Ax=A((VA-I)x+v)+AVy$ and

$\widehat{1^Tx}(z)-1^Tx=(w,Ax)+(w,y)+\varrho-(1,x)=(A^Tw-1,x)+\varrho+(w,y)$.

We therefore obtain

$E\|\hat{x}(z)-x\|^2= E\{\|(VA-I)x+v\|^2+\|Vy\|^2+2((VA-I)x+v,Vy)\}$ ,

$E\|A\hat{x}(z)-Ax\|^2 = E\{\|A((VA-I)x+v)\|^2+\|AVy\|^2+2(A((VA-I)x+v),AVy)\}$ and

$E(\widehat{1^Tx}(z)-1^Tx)^2 = E\{((A^Tw-1,x)+\varrho)^2+(w,y)^2+2((A^Tw-1,x)+\varrho)(w,y)\}$ .

The first term in each of the above written sums within braces is nonrandom. Hence, the operation E(.) does not alter these terms. Mathematical expectation of the second terms is verified by using the formulae (5.11) through (5.13) of Lemma 5.4. Since $\eta$ is a centralized random vector, we furthermore obtain $E(w,y)=0$ due to $E(V^T((VA-I)x+v),y)=0$ and $E(V^TA^TA((VA-I)x+v),y)=0$. Thus, the third terms vanish. #

The formulae (5.17)-(5.19) will be applied for further studies below. We are interested in determining linear estimators with small mean square errors. However, beforehand we take into consideration an additional feature of estimators.

Definition 5.7:

Estimators $\hat{x}(z)$ and $\widehat{1^Tx}(z)$ for the vector reconstruction from the points of view of identification and control and for the functional reconstruction in the linear model (5.1) are said to be unbiased if, independently of the present vector $x\in R^n$,

$$E\ \hat{x}(z) = x\ , \tag{5.20}$$

$$E\ A\ \hat{x}(z) = A\ x \tag{5.21}$$

and

$$E\, \widehat{1^T x}(z) = 1^T x\ , \tag{5.22}$$

respectively. o

Unbiased estimators guarantee that there is a mean coincidence of estimates thus obtained with the true parameter vector or with the true functional value to be reconstructed.

Definition 5.8:

Supposed that linear unbiased estimators $\hat{x}(z)$ and $\widehat{1^T x}(z)$ satisfying (5.20), (5.21) or (5.22) exist, the estimators $\hat{x}_{blu1}(z)$, $\hat{x}_{blu2}(z)$ and $(\widehat{1^T x})_{blu}(z)$ are called the best linear unbiased estimators for the vector reconstruction from the point of view of identification and control and for the functional reconstruction if they are minimizers of the mean square error risk (5.5), (5.7) and (5.6) in the class of linear unbiased estimators. o

Best linear unbiased estimators are unbiased estimators with a minimum sum of error component variances. Questions concerning the existence and construction of best estimators $\hat{x}_{blu1}(z)$, $\hat{x}_{blu2}(z)$ and $(\widehat{1^T x})_{blu}(z)$ are answered by the following theorem.

Theorem 5.9:

Linear estimators satisfy the conditions (5.20), (5.21) and (5.22) respectively, if and only if

$$V\cdot A = I\ ,\quad v = 0 \qquad \text{for (5.20)}\ , \tag{5.23}$$

$$A\cdot V\cdot A = A\ ,\quad A\cdot v = 0 \quad \text{for (5.21)} \tag{5.24}$$

and

$$A^T\cdot w = 1\ ,\quad \varrho = 0 \qquad \text{for (5.22)}\ . \tag{5.25}$$

Matrices V satisfying (5.23) exist in the full-rank case rank(A)=n, whereas matrices V according to (5.24) always exist. The existence of a vector w satisfying (5.25) is ensured independently of the special structure of 1 if rank(A)=n, but also if rank(A)<n and $1\in\mathbb{R}(A^T)$, i.e., if 1 is orthogonal to the null-space of A. Provided that linear unbiased estimators exist, the best linear estimators have the form

$$\hat{x}_{blu1}(z) = (A^T C^{-1} A)^{-1} A^T C^{-1} z\ , \tag{5.26}$$

$$\hat{x}_{blu2}(z) = (A^T C^{-1} A)^{+} A^T C^{-1} z + v_N = (C^{-1/2} A)^{+} C^{-1/2} z + v_N\ ,$$

$$v_N \in \mathbb{N}(A) \text{ arbitrarily chosen}\ , \tag{5.27}$$

and

$$(\widehat{1^T x})_{blu}(z) = 1^T (A^T C^{-1} A)^{+} A^T C^{-1} z\ .\ \circ \tag{5.28}$$

Proof:

The assertions of this theorem regarding the solvability of VA=I, AVA=A and $A^T w=1$ are well-known from linear algebra (see e.g. ZIELKE [489]). Note that AVA=A is one of the Penrose equations satisfied for all generalized inverses $V=A^-$ (see RAO [362]). Now suppose $\mathrm{rank}(A)=n$ and therefore $\mathrm{rank}(A^TC^{-1}A)=n$ . Then, $V=(A^TC^{-1}A)^{-1}A^TC^{-1}$ obviously satisfies $VA=I$. Hence, (5.26) is a linear unbiased estimator. By formula (5.17) we obtain

$$E\,\|\hat{x}_{blu1}(z)-x\|^2 = \mathrm{trace}((A^TC^{-1}A)^{-1}) \; . \tag{5.29}$$

Now consider an arbitrarily chosen matrix $\tilde{V}\in\mathbb{R}^{n\times m}$ satisfying $\tilde{V}A=0$. We can state the equations $\mathrm{trace}(((A^TC^{-1}A)^{-1}A^TC^{-1}+\tilde{V})C(C^{-1}A(A^TC^{-1}A)^{-1}+\tilde{V}^T)) = \mathrm{trace}((A^TC^{-1}A)^{-1})+ 2\mathrm{trace}(\tilde{V}A(A^TC^{-1}A)^{-1}) + \mathrm{trace}(\tilde{V}C\tilde{V}^T) = \mathrm{trace}((A^TC^{-1}A)^{-1})+ \mathrm{trace}(\tilde{V}C\tilde{V}^T)$. Owing to (5.16), $\mathrm{trace}(\tilde{V}C\tilde{V}^T)= \mathrm{trace}((\tilde{V}C^{1/2})(\tilde{V}C^{1/2})^T)\geq 0$ and $\mathrm{trace}(\tilde{V}C\tilde{V}^T)= 0$ iff $\tilde{V}C^{1/2}=0$, i.e., if $\tilde{V}=0$. Hence, (5.26) is the uniquely determined minimizer of (5.5) in the class of linear unbiased estimators. Considering $\hat{x}_{blu2}(z)$ and $(\widehat{1^Tx})_{blu}(z)$ according to the formulae (5.27) and (5.28), we derive the unbiasedness from the obviously valid equations $\tilde{A}(\tilde{A}^T\tilde{A})^+\tilde{A}^T\tilde{A}=\tilde{A}$ and $\tilde{A}^T\tilde{A}(\tilde{A}^T\tilde{A})^+1=1$ with $1\in\mathcal{R}(A^T)$ by setting $\tilde{A}:=C^{-1/2}A$. Moreover, one obtains

$$E\,\|\hat{x}_{blu2}(z)-x\|^2 = \mathrm{trace}(A(A^TC^{-1}A)^+A^T) \tag{5.30}$$

and

$$E((\widehat{1^Tx})_{blu}(z)-1^Tx)^2 = ((A^TC^{-1}A)^+1,1) \; . \tag{5.31}$$

As we have shown in the proof concerning (5.26), it may be seen that terms added to (5.27) and (5.28) never reduce the error risk (5.7) and (5.6). All best linear unbiased estimators are uniquely determined with the exception of an arbitrarily chosen null-space term $x_N$ in $\hat{x}_{blu2}(z)$. #

There are various different reasons to give up assuming the unbiasedness of considered estimators in the linear model (see also Sec. 5.1.2.). If we deal with biased estimators, then the first terms in the formulae (5.17)-(5.19) may considerably contribute to the total errors of estimation. In such a case, minimizers of (5.5)-(5.7) are not of interest since they generally fail to be independent of the unknown vector x. However, one can prescribe upper bounds on the Euclidean norm of x and ask for estimators that minimize the maximum value of mean square risks over all admissible vectors. These so-called minimax estimators are frequently difficult to verify (see HUMAK [217]). We present minimax estimators for the vector and functional reconstruction problems from the point of view of identification.

Definition 5.10:
For given $c>0$, the estimators $\hat{x}_{lmm}(z)$ and $(\widehat{1^Tx})_{lmm}(z)$ are called the linear minimax estimators if they are minimizers of

$$\max_{\|x\|^2 \leq c} E \|\hat{x}-x\|^2 \quad (5.32)$$

and

$$\max_{\|x\|^2 \leq c} E (\widehat{1^T x}(z)-1^Tx)^2 , \quad (5.33)$$

respectively, in the class of linear estimators (5.3), (5.4). ○

To establish linear minimax estimators, we explicitly express the maxima of the terms (5.17) and (5.18).

Lemma 5.11:
If $M \in R^{k \times n}$, then

$$\min_{u \in R^k} \max_{\|x\|^2 \leq c} \|Mx+u\|^2 = c \cdot \|M\|^2 \quad . \circ \quad (5.34)$$

Proof:
It is evident that $\max_{\|x\|^2 \leq c} \|Mx\|^2 = c \cdot \|M\|^2$. Now assume to have a vector $0 \neq u \in R^k$ with $\max_{\|x\|^2 \leq c} \|Mx+u\|^2 \leq c \cdot \|M\|^2$. Due to $\|Mx+u\|^2 = \|Mx\|^2 + 2(Mx,u)+\|u\|^2$, the above assumption implies $(Mx^{(0)},u)<0$ for all vectors $x^{(0)}$ with $\|x^{(0)}\|^2 \leq c$ and $\|Mx^{(0)}\|^2 = c \cdot \|M\|^2$. This contradicts the fact that $\|M(-x^{(0)})\|^2 = c \cdot \|M\|^2$ holds whenever $\|Mx^{(0)}\|^2 = c \cdot \|M\|^2$ is provided. Hence, either $(Mx^{(0)},u)$ or $(M(-x^{(0)}),u)$ is nonnegative. This proves the lemma. #

From Lemma 5.11 we can derive that linear minimax estimators under consideration are homogeneous linear transformations, i.e., $v=0$ and $\varrho=0$. In the vector reconstruction case, Lemma 5.11 applies with $k=m$, whereas $k=1$ is of interest in the functional reconstruction case. If V denotes an arbitrarily chosen matrix $V \in R^{n \times m}$, we obtain

$$\min_{V \in R^{n \times m}} \max_{\|x\|^2 \leq c} E \|Vz-x\|^2 = \min_{V \in R^{n \times m}} \left\{ c\|VA-I\|^2 + \operatorname{trace}(VCV^T) \right\} .$$

Moreover, if we have a vector w that arbitrarily varies in $R^m$, then

$$\min_{w \in R^m} \max_{\|x\|^2 \leq c} E (w^Tz-1^Tx)^2 = \min_{w \in R^m} \left\{ c\|A^Tw-1\|^2 + (Cw,w) \right\} .$$

Lemma 5.12:
The matrix

$$V_{lmm} = \begin{cases} \dfrac{c}{c+\operatorname{trace}((A^TC^{-1}A)^{-1})} \cdot (A^TC^{-1}A)^{-1}A^TC^{-1} & \text{if } \operatorname{rank}(A)= \\ 0 & \text{if } \operatorname{rank}(A)< \end{cases} \quad (5.35)$$

is the uniquely determined solution to the matrix optimization problem

$$\underset{V\in R^{n\times m}}{\text{minimize}} \left\{ c\|VA-I\|^2+\text{trace}(VCV^T) \right\} . \circ \qquad (5.36)$$

Proof: FRIEDRICH [135], [138, p.35].

Theorem 5.13:
The linear minimax estimators $\hat{x}_{lmm}(z)$ and $\widehat{(1^Tx)}_{lmm}(z)$ are uniquely determined by the formulae

$$\hat{x}_{lmm}(z) = \frac{c}{c+\text{trace}((A^TC^{-1}A)^{-1})}\cdot(A^TC^{-1}A)^{-1}A^TC^{-1}z \quad \text{if rank}(A)=n, \qquad (5.37)$$

$$\hat{x}_{lmm}(z) = 0 \quad \text{if rank}(A)<n \qquad (5.38)$$

and

$$\widehat{(1^Tx)}_{lmm}(z) = 1^T(A^TC^{-1}A+\frac{1}{c}\cdot I)^{-1}A^TC^{-1}z = 1^TA^T(AA^T+\frac{1}{c}\cdot C)^{-1}z. \qquad (5.39)$$

$\circ$

Proof:
Uniqueness and form of $\hat{x}_{lmm}(z)$ result from Lemma 5.12. Now consider the functional case, where $c\|A^Tw-1\|^2+(Cw,w)$ is to be minimized with respect to $w\in R^m$. Evidently, we obtain a quadratic form $((c\cdot AA^T+C)w,w)-2(w,c\cdot A1)+c\|1\|^2$, the unique minimizer of which is a solution to the system of linear algebraic equations $(c\cdot AA^T+C)\,w = c\cdot A1$ (see Lemma 4.32). Since $c\cdot AA^T+C$ is a positive definite, regular matrix, this minimizer is equal to $w_{lmm}=(AA^T+\frac{1}{c}\cdot C)^{-1}A1$ . In view of the matrix identity

$$(A^TC^{-1}A+M^{-1})^{-1}A^TC^{-1} = MA^T(AMA^T+C)^{-1} \qquad (5.40)$$

that is valid for all regular quadratic matrices $M\in R^{n\times n}$, $C\in R^{m\times m}$ and an arbitrary matrix $A\in R^{m\times n}$, we find the second form $w_{lmm}= C^{-1}A(A^TC^{-1}A+\frac{1}{c}\cdot I)^{-1}1$ by setting $M:=c\cdot I$ .

Note that

$$\max_{\|x\|^2\leq c} E\,\|\hat{x}_{lmm}(z)-x\|^2 = \begin{cases} \frac{c}{c+\text{trace}((A^TC^{-1}A)^{-1})}\cdot\text{trace}((A^TC^{-1}A)^{-1}) & \text{if rank}(A)=n \\ c & \text{if rank}(A)<n \end{cases} \qquad (5.41)$$

and

$$\max_{\|x\|^2\leq c} E\,(\widehat{(1^Tx)}_{lmm}(z)-1^Tx)^2 = ((A^TC^{-1}A+\frac{1}{c}\cdot I)1,1) . \# \qquad (5.42)$$

Remark 5.14:
Permitting a bias $\|VA-I\|^2>0$ and $\|A^Tw-1\|^2>0$ can be a successful way to reduce the total error of estimation. Namely, for rank(A)=n,

$$\max_{\|x\|^2\leq c} E\,\|\hat{x}_{lmm}(z)-x\|^2 = \frac{c}{c+\text{trace}((A^TC^{-1}A)^{-1})}\cdot E\,\|\hat{x}_{blu1}(z)-x\|^2 .$$

The damping factor $\frac{c}{c+\mathrm{trace}((A^TC^{-1}A)^{-1})}<1$ grows with c. Therefore, the higher the degree of objective a priori information, i.e., the smaller the value c, the smaller the worst case error of minimax estimates becomes. The same effect arises in the functional reconstruction case. Here, for $c>0$ and $0\neq l\in\mathcal{R}(A^T)$, $((A^TC^{-1}A+\frac{1}{c}\cdot I)^{-1}l,l)<$ $<((A^TC^{-1}A)^{+}l,l)$. The difference between both sides of the above inequality is increasing if c is decreasing. Owing to Theorem 5.13 it can also be noticed that the more informative functional reconstruction case leads to a unified structure of linear minimax estimators, independently of the matrix rank of A. The less informative vector reconstruction case, however, has to distinguish between the full-rank situation, $\mathrm{rank}(A)=n$, where the linear minimax estimator $\hat{x}_{lmm}(z)$ is a shrunken estimator, i.e., a damped modification of $\hat{x}_{blu1}(z)$ with a damping factor $\frac{c}{c+\mathrm{trace}((A^TC^{-1}A)^{-1})}$, and the deficient-rank situation, $\mathrm{rank}(A)<n$, where the trivial estimator $\hat{x}_{lmm}(z)=0$ indicates the associated insufficiency of the data for the reconstruction of the whole unknown vector. Finally, note that $(\widehat{l^Tx})_{lmm}(z)=l^T\cdot\hat{x}(z)$ with $\hat{x}(z)$ being a representative of the class of ridge estimators $\hat{x}_{ridge}(z)$ that we are going to study in the coming section. o

5.1.2. Ridge Estimators and Cross-Validation

In Sec. 5.1.1., we have avoided taking into consideration the stability of estimates with respect to perturbations in the data vector z. The stability corresponds to small values of the spectral norm $\|V\|$ if linear estimators (5.3) are used. Namley, we have

$$\|\hat{x}(z^{(1)})-\hat{x}(z^{(2)})\|\leq\|V\|\cdot\|z^{(1)}-z^{(2)}\| \,. \tag{5.43}$$

It is evident from (5.26) and (5.27) that the matrix norm $\|V\|$ associated with $\hat{x}_{blu1}(z)$ and $\hat{x}_{blu2}(z)$ tends to infinity as the smallest singular value of the matrix $C^{-1/2}A$ tends to zero. Therefore, ill-conditioned linear models (5.1) with large values $\|A^+\|$ are related to instable best linear unbiased estimators.

Best linear unbiased estimators are sometimes called the least squares estimators. They represent minimizers of the norm square functional $\|\tilde{A}x-\tilde{z}\|^2$ with $\tilde{A}=C^{-1/2}A$ and $\tilde{z}=C^{-1/2}z$. An aim of Chapter 4 was to replace least squares solutions $x_{ls}$ by regularized solutions $x_{(\alpha)}$. In an analogous way we can approximate least squares estimators by biased estimators in order to achieve a higher degree of stability.

Definition 5.15:

The estimator

$$\hat{x}_{gridge}(z) := (A^TC^{-1}A+M^{-1})^{-1}A^TC^{-1}z = MA^T(AMA^T+C)^{-1}z \qquad (5.44)$$

(cf. formula (5.40)) is said to be a generalized ridge estimator depending on the positive definite symmetric ridge parameter-matrix $M \in R^{n\times n}$. The important special case $M = \frac{1}{\alpha}\cdot I$ with

$$\hat{x}_{ridge}(z) := (A^TC^{-1}A+\alpha\cdot I)^{-1}A^TC^{-1}z = A^T(AA^T+\alpha\cdot C)^{-1}z \ , \ \alpha>0 \qquad (5.45)$$

represents the ordinary form of ridge estimators. The value $\alpha>0$ is termed the ridge parameter. o

Remark 5.16:

If the observation error components are uncorrelated and the variance $\sigma^2$ is common to all components, i.e.,

$$C = \sigma^2\cdot I \ , \qquad (5.46)$$

then $\hat{x}_{ridge}(z)$ provides estimates that coincide with regularized solutions $x_{(\alpha)}$. Thus, the vector families defined by the formulae (5.45) and (4.80), (4.82) with $\Phi = I$, $\bar{x} = 0$ are identical. o

Many authors have dealt with ordinary and generalized ridge estimators (see e.g. HOERL/KENNARD [191-92], GOLUB/HEATH/WAHBA [159], MARQUARDT [292], TAUTENHAHN [423], MITCHELL/DRAPER [307] and HUA/GUNST [216]). It can be shown that the norm square error risk of ridge estimators

$$E\|\hat{x}_{ridge}(z)-x\|^2 = \alpha^2\|(A^TC^{-1}A+\alpha\cdot I)^{-1}x\|^2 + \text{trace}\{(A^TC^{-1}A+\alpha\cdot I)^{-1}\cdot \cdot A^TC^{-1}A(A^TC^{-1}A+\alpha\cdot I)^{-1}\} \ , \ \alpha>0 \qquad (5.47)$$

gets smaller than the mean square error risk $E\|\hat{x}_{blu1}(z)-x\|^2$ (see (5.29)) whenever rank(A)=n and $\alpha=\alpha(x)$ is appropriately chosen in dependence of the true vector x. Namely, based on singular values $\tilde{\sigma}_1\geq\tilde{\sigma}_2\geq\ldots\geq\tilde{\sigma}_n>0$ of $C^{-1/2}A$ we can point out that

$$E\|\hat{x}_{ridge}(z)-x\|^2 = \sum_{i=1}^{n} \frac{\alpha^2(x,\tilde{u}^{(i)})^2+\tilde{\sigma}_i^2}{(\tilde{\sigma}_i^2+\alpha)^2} \qquad (5.48)$$

and

$$E\|\hat{x}_{blu1}(z)-x\|^2 = \sum_{i=1}^{n} \frac{1}{\tilde{\sigma}_i^2} \ . \qquad (5.49)$$

The vectors $\tilde{u}^{(1)},\ldots,\tilde{u}^{(n)}$, in this context, form the columns of the orthogonal matrix $\tilde{U}$ in the singular value decomposition

$$C^{-\frac{1}{2}}\cdot A = \tilde{V}\cdot\begin{pmatrix} \tilde{\sigma}_1 & & 0 \\ & \tilde{\sigma}_2 & \\ 0 & & \ddots \ \tilde{\sigma}_n \\ \hline & 0 & \end{pmatrix}\cdot\tilde{U}^T \qquad \text{(compare the formula (4.78))}.$$

**Theorem 5.17:**
If rank(A)=n, then there is, for any vector $x \in R^n$, a ridge parameter $\alpha>0$ depending on x such that

$$E\|\hat{x}_{ridge}(z)-x\|^2 < E\|\hat{x}_{blu1}(z)-x\|^2 . \qquad (5.50)$$

**Proof:**
If x=0, then the inequality $\sum_{i=1}^{n} \frac{\tilde{\sigma}_i^2}{(\tilde{\sigma}_i^2+\alpha)^2} < \sum_{i=1}^{n} \frac{1}{\tilde{\sigma}_i^2}$ yields (5.50) for all $\alpha>0$. If otherwise $x \neq 0$, then

$$0<\alpha<\frac{2}{\max\limits_{1\le i\le n} (x,\tilde{u}^{(i)})^2} \text{ implies } \frac{\alpha^2(x,\tilde{u}^{(i)})^2+\tilde{\sigma}_i^2}{(\tilde{\sigma}_i^2+\alpha)^2} < \frac{1}{\tilde{\sigma}_i^2} \quad \text{for all } 1\le i\le n.$$

Due to (5.48) and (5.49) this completes the proof. #

There is no ridge estimator with fixed $\alpha>0$ or with a fixed parameter matrix M that estimates, for all $x \in R^n$, uniformly better than the best linear unbiased estimator. As for the uniform improvement of least squares estimators using Stein-type estimations we refer to O. BUNKE [44] (see also [416] and [423]).

The principal coincidence of ridge estimation and Tikhonov's regularization is refelcted by a principal coincidence of the problems which arise from the choice of ridge and regularization parameters. In addition to the parameter choice criteria of Sec. 4.1., stochastically motivated criteria have been developed for selecting a good ridge parameter. For simplicity, we are going to outline some of these criteria under the assumption of formula (5.46).

A first ridge fashion of Morozov's discrepancy principle

$$\alpha_{discr1} : \quad \|A\,\hat{x}_{ridge}(z) - z\|^2 = \varrho \cdot m \cdot \sigma^2 \qquad (5.51)$$

with $E\|y\|^2 = m\cdot\sigma^2$ and a fudge factor $0<\varrho\le 1$ is proposed e.g. by WAHBA/WOLD [477]. Simulation experiments have suggested choosing a value $\varrho$ of the interval $0.7\le\varrho\le 0.95$ for the fudge factor. The introduction of $\varrho$ is aimed at overcoming a tendency of the discrepancy principle in regularization towards the selection of oversmoothed solutions.

Another technique to avoid this oversmoothing is based on the second ridge variant of the discrepancy principle, viz. (see [33]):

$$\alpha_{discr2} : \quad \|A\,\hat{x}_{ridge}(z) - z\|^2 + \text{trace}(A(A^TA+\alpha\cdot I)^{-1}A^T) = m\cdot\sigma^2 . \qquad (5.52)$$

A promising method of determining $\alpha$ from data information, the method of generalized cross-validation, has been developed by

WAHBA et al. (see [467-68], [470-71], [473], [475-77], [77] and [159]). The associated ridge parameter $\alpha_{gcv}$ is a minimizer of

$$\alpha_{gcv} : \underset{\alpha>0}{\text{minimize}} \; \frac{\|(AA^T+\alpha\cdot I)^{-1}\cdot z\|^2}{(\text{trace}(A^TA+\alpha\cdot I)^{-1})^2} \; . \qquad (5.53)$$

In the paper [159], GOLUB et al. gave a deep mathematical and statistical insight into the mechanism of generalized cross-validation methods, their motivation and applicability. The efficient computational solution of the search problem (5.53) is discussed in [33]. As for the iterative application of this method to constrained problems, we refer to Sec. 6.4. of Wahba's report [470]. Numerical aspects of realizing cross-validation techniques that may differ from (5.53) are studied in [138, p.68] (cf. also CLARK [69] and KOCOTOV [248]).

### 5.1.3. The Reduction Problem

In this section, we will briefly discuss a problem stated by PYTEV [358-59], which he named the reduction problem. Homogeneous linear estimators $\hat{x}(z) = Vz$ , $V\in R^{n\times m}$ , applied to linear modelling (5.1) and (5.2), are the objects of interest here. The reduction problem is aimed at minimizing the bias of the estimator provided that the mean square influence of random observation errors gets limited.

In the following, we formulate the reduction problem for the identification and for the control aspect of the vector reconstruction case each in its own manner. To express the bias in the sequel, we will use the Frobenius norm, which is defined as

$$\|A\|_F := (\text{trace}(A^TA))^{1/2} = \Big( \sum_{\substack{i=1(1)m \\ j=1(1)n}} \sum a_{ij}^2 \Big)^{1/2} \text{ if } A=(a_{ij})_{j=1(1)n}^{i=1(1)m} \; . \qquad (5.54)$$

Definition 5.18:

For given $\varepsilon>0$, the estimator $\hat{x}_{red1}(z)= V_{red1}\cdot z$ concerning the identification problem and the estimator $\hat{x}_{red2}(z)= V_{red2}\cdot z$ concerning the control problem are called the reduction estimators if the matrices $V_{red1}\in R^{n\times m}$ and $V_{red2}\in R^{n\times m}$ are solutions to the problems

$$\underset{\substack{V\in R^{n\times m} \\ E\,\|Vy\|^2 \leq \varepsilon}}{\text{minimize}} \|VA-I\|_F^2 , \quad E\,\|Vy\|^2 = \text{trace}(VCV^T) \qquad (5.55)$$

and

$$\underset{\substack{V\in R^{n\times m} \\ E\,\|AVy\|^2 \leq \varepsilon}}{\text{minimize}} \|AVA-A\|_F^2 , \quad E\,\|AVy\|^2 = \text{trace}(AVCV^TA^T) , \qquad (5.56)$$

respectively. ○

A Lagrange multiplier approach as already used to analyse Tikhonov's regularization method (see Sec. 4.1.2.) allows us to verify the matrices $V_{red1}$ and $V_{red2}$. Consider, for given $\alpha>0$, the convex functionals

$$F_{red1}(V,\alpha) := \|VA-I\|_F^2+\alpha\cdot\|VC^{1/2}\|_F^2 = = \operatorname{trace}\{(VA-I)(VA-I)^T+\alpha\cdot VCV^T\} \tag{5.57}$$

and

$$F_{red2}(V,\alpha) := \|AVA-A\|_F^2+\alpha\cdot\|AVC^{1/2}\|_F^2 = = \operatorname{trace}\{A(VA-I)(VA-I)^TA^T+\alpha\cdot AVCV^TA^T\}. \tag{5.58}$$

Lemma 5.19:

The matrix $V_{min}\in R^{n\times m}$ is a solution to the matrix optimization problem

$$\underset{V\in R^{n\times m}}{\text{minimize}} \quad \operatorname{trace}\{G((VA-I)B(VA-I)^T+VCV^T)G^T\} \tag{5.59}$$

with given matrices $G\in R^{k\times n}$, $A\in R^{m\times n}$ and given symmetric positive definite matrices $B\in R^{n\times n}$, $C\in R^{m\times m}$ if and only if

$$V_{min} = BA^T(ABA^T+C)^{-1}+V_N = (A^TC^{-1}A+B^{-1})^{-1}A^TC^{-1}+V_N \tag{5.60}$$

and $V_N\in R^{n\times m}$ is an arbitrarily chosen matrix satisfying the equation

$$G\cdot V_N = 0 \,. \tag{5.61}$$

Proof:

Let be $V=(A^TC^{-1}A+B^{-1})^{-1}A^TC^{-1}+\tilde{V}$, where $\tilde{V}\in R^{n\times m}$ is an arbitrarily chosen matrix. Then, one can verify $\operatorname{trace}\{G((VA-I)B(VA-I)^T+VCV^T)G^T\} = \operatorname{trace}\{G(A^TC^{-1}A+B^{-1})^{-1}G^T\}+ + \operatorname{trace}\{G\tilde{V}(ABA^T+C)\tilde{V}^TG^T\}$ with

$$\operatorname{trace}\{G((V_{min}A-I)B(V_{min}A-I)^T+V_{min}CV_{min}^T)G^T\} = =\operatorname{trace}\{G(A^TC^{-1}A+B^{-1})^{-1}G^T\} \,. \tag{5.62}$$

However, the nonnegative term trace $G\tilde{V}(ABA^T+C)\tilde{V}^TG^T$ equals zero iff $G\tilde{V}(ABA^T+C)^{1/2}= 0$ and thus $G\tilde{V} = 0$. Due to (5.40) the two different forms of $V_{min}$ in formula (5.60) are equivalent. #

Owing to Lemma 5.19, the functional (5.57) has a unique minimizer

$$V_1(\alpha) = A^T(AA^T+\alpha\cdot C)^{-1} = (A^TC^{-1}A+\alpha\cdot I)^{-1}A^TC^{-1} \tag{5.63}$$

obtained by substituting $k:=n$, $G:=I$ and $B:=\frac{1}{\alpha}\cdot I$. The same lemma yields a representation for the minimizer of (5.58) as

$$V_2(\alpha) = V_1(\alpha) + V_N \,, \quad AV_N = 0 \,, \tag{5.64}$$

where $k:=m$, $G:=A$ and $B:=\frac{1}{\alpha}\cdot I$ .

In the sequel, let us exclude the pathological case $A=0$ . Then, the coming theorem gives a complete characterization of the reduction estimators.

Theorem 5.20:

Provided $A\neq 0$, $\mathcal{E}_1 := \operatorname{trace}((A^TC^{-1}A)^+)$ and $\mathcal{E}_2 := \operatorname{trace}(A(A^TC^{-1}A)^+A^T)$, we have

$$V_{red1} = \begin{cases} V_1(\alpha) & \text{if } 0<\mathcal{E}<\mathcal{E}_1 \\ (C^{-1/2}A)^+C^{-1/2} = (A^TC^{-1}A)^+A^TC^{-1} & \text{if } \mathcal{E}\geq\mathcal{E}_1 \\ 0 & \text{if } \mathcal{E}=0\,, \end{cases} \tag{5.65}$$

with the matrix $V_1(\alpha)$ defined by formula (5.63), and

$$V_{red2} = \begin{cases} V_2(\alpha) = V_1(\alpha) + V_N & \text{if } 0<\mathcal{E}<\mathcal{E}_2 \\ (C^{-1/2}A)^+C^{-1/2}+V_N = (A^TC^{-1}A)^+A^TC^{-1}+V_N & \text{if } \mathcal{E}\geq\mathcal{E}_2 \\ V_N & \text{if } \mathcal{E}=0\,, \end{cases} \tag{5.66}$$

where $V_N\in\mathbb{R}^{n\times m}$ is an arbitrary matrix satisfying $AV_N=0$. The value $\alpha=\alpha(\mathcal{E})$ in (5.65) and (5.66) represents the uniquely determined solution to the equation

$$\operatorname{trace}(V_1(\alpha)C(V_1(\alpha))^T) = \mathcal{E} \tag{5.67}$$

and to the equation

$$\operatorname{trace}(AV_2(\alpha)C(V_2(\alpha))^TA^T) = \mathcal{E}\,, \tag{5.68}$$

respectively. o

Proof: PYTEV [359, §4].

The assertions of this theorem become clear if the problems (5.55), (5.56) and the functionals (5.57), (5.58) are considered to be matrix generalizations to the problem (4.7) and to the functional (4.1) in the linear unconstrained case. If we study the functions of $\alpha$ , $\|V_1(\alpha)A-I\|_F$ (or $\|AV_1(\alpha)A-A\|_F$) and $\|V_1(\alpha)C^{1/2}\|_F^2$ (or $\|AV_1(\alpha)C^{1/2}\|_F^2$ ), then the monotonicity and continuity of functions $s_{res}(\alpha)$ and $s_{\Omega}(\alpha)$ (see Theorem 4.38) carry over to the corresponding function considered here. The limits (4.96), (4.97) arrive at the following forms:

$$\left.\begin{aligned} &\lim_{\alpha\to 0+0} \|V_1(\alpha)A-I\|_F = \|A^+A-I\|_F\,, \\ &\lim_{\alpha\to 0+0} \|AV_1(\alpha)A-A\|_F = 0\,, \\ &\lim_{\alpha\to 0+0} \|V_1(\alpha)C^{1/2}\|_F^2 = \operatorname{trace}((A^TC^{-1}A)^+) = \mathcal{E}_1\,, \\ &\lim_{\alpha\to 0+0} \|AV_1(\alpha)C^{1/2}\|_F^2 = \operatorname{trace}(A(A^TC^{-1}A)^+A^T) = \mathcal{E}_2 \end{aligned}\right\} \tag{5.69}$$

and

$$\left.\begin{aligned}&\lim_{\alpha\to\infty}\|V_1(\alpha)A-I\|_F = \|I\|_F = \sqrt{n}\ ,\\&\lim_{\alpha\to\infty}\|AV_1(\alpha)A-A\|_F = \|A\|_F\ ,\\&\lim_{\alpha\to\infty}\|V_1(\alpha)C^{1/2}\|_F^2 = \lim_{\alpha\to\infty}\|AV_1(\alpha)C^{1/2}\|_F^2 = 0\ .\end{aligned}\right\}\quad(5.70)$$

If taking into account the above limits, then the unique solvability of Eqs. (5.67) and (5.68), for $0<\varepsilon<\varepsilon_1$ and for $0<\varepsilon<\varepsilon_2$, respectively, is a consequence of the continuity with respect to $\alpha$ of all terms forming the functionals $F_{red1}$ and $F_{red2}$. #

The given brief study of the reduction problem has yielded an interpretation of ridge estimators that appear in the form $\hat{x}_{red1}(z)= V_1(\alpha)\cdot z$ if $0<\varepsilon<\varepsilon_1$ and $\hat{x}_{red2}(z)= V_1(\alpha)\cdot z + v_N$, $v_N\in \mathbb{N}(A)$, if $0<\varepsilon<\varepsilon_2$. Whenever the constraint $E\|VC^{1/2}\|_F^2\leq\varepsilon$ (or $E\|AVC^{1/2}\|_F^2\leq\varepsilon$) is not active, i.e., $\varepsilon\geq\varepsilon_1$ (or $\varepsilon\geq\varepsilon_2$), then the estimator $\hat{x}_{blu2}(z)$ is a reduction estimator. This gives a further plausible explanation for the name least squares estimator of the best linear unbiased estimator in that case.

Finally, note that the converse problems

$$\begin{aligned}&\text{minimize}\quad E\,\|Vy\|^2\\&V\in R^{n\times m}\\&\|VA-I\|_F^2\leq\varepsilon\end{aligned}\qquad(5.71)$$

and

$$\begin{aligned}&\text{minimize}\quad E\,\|AVy\|^2\\&V\in R^{n\times m}\\&\|AVA-A\|_F^2\leq\varepsilon\end{aligned}\quad ,\qquad(5.72)$$

which are aimed at finding the minimum variance estimators when bounds for the bias are prescribed, can be treated analogously by the Lagrange multiplier approach studied in Chap. 4. Optimal estimators will again be recruited from the class of ridge and least squares estimators. It is evident that a minimizer $V_1(\alpha)$ of (5.57) satisfying the equation $\|V_1(\alpha)A-I\|_F^2=\varepsilon$ is also a minimizer of (5.71). An analogous assertion can be stated for the relationship regarding $V_2(\alpha)$, (5.58) and (5.72).

## 5.1.4. Remarks on the Nonlinear Model

We are now going to complete the study of the non-Bayesian stochastic approach by some remarks on the nonlinear model. Instead of (5.1) let us consider the nonlinear model

$$A\,x + y = z\ ,\quad A:\ R^n\to R^m \text{ nonlinear},\ x\in R^n,\ y,z\in R^m,\qquad(5.73)$$

where the vector $x$ is fixed but unknown, and $y$ again represents a realization of a random vector $\eta$ subject to (5.2). In statistics, the problem of determining x from z by using an estimator $\hat{x}(z)$ is a nonlinear regression problem. As for the current state of knowledge of nonlinear models, we refer to the exhaustive treatise of the subject in the book of HUMAK [218]. However, if one seeks an introduction to nonlinear regression and to associated numerical problems, the paper [42] of H. BUNKE can be recommended.

In working with the vector reconstruction problem, the mean square error of estimates obtained by a homogeneous linear estimator $\hat{x}(z)= Vz$, $V\in R^{n\times m}$, is described by the formula

$$E\|\hat{x}(z)-x\|^2 = \|VAx-x\|^2 + \mathrm{trace}(VCV^T) \tag{5.74}$$

(cf. formula (5.17)). Under the assumption (5.46), $C= \sigma^2 I$, this error is minimized, for given x, if $V:=V_{min}(x)$ has the form

$$V_{min}(x) = \frac{x\cdot(Ax)^T}{\|Ax\|^2+ \sigma^2} \quad . \tag{5.75}$$

With $\hat{x}_{min}(z):= V_{min}(x)\cdot z$ we have

$$E\|\hat{x}_{min}(z)-x\|^2 = \sigma^2 \cdot \frac{\|x\|^2}{\|Ax\|^2+ \sigma^2} \quad . \tag{5.76}$$

For the proof,see [138, p.74]. The formula (5.76) indicates that best estimation results are possible if the quotient $\|x\| / \|Ax\|$ is small. In the linear model, this would agree with a dominance of principal components in the solution vector x. Since we do not know x, the estimator $\hat{x}_{min}(z)$ cannot be exploited directly. However, this discussion provides a new evaluation functional

$$\Omega(x) = \frac{\|x\|^2}{\|Ax\|^2 + \sigma^2} \tag{5.77}$$

for the method of Tikhonov regularization. By using (5.77) the regularization technique selects representatives of the set of data-compatible vectors, which are estimable fairly well. This is a stochastically motivated alternative to the predomiantly used smoothing functionals $\Omega(x)$ (cf. formula (2.95)).

## 5.2. The Bayesian Case

### 5.2.1. Statistical Regularization

The Bayesian model in its general finite dimensional form is a modification to (5.1) in the linear and (5.73) in the nonlinear case. It is based on a randomized solution vector x. We are going to consider the Bayesian noisy data problem (see Def. 2.65) as follows:

$$A\,x + y = z\ , \quad A: R^n \to R^m,\ x \in R^n,\ y, z \in R^m \ . \tag{5.78}$$

Both the solution vector x and the perturbation vector y are realizations of random vectors $\xi \in R^n$ and $\eta \in R^m$, respectively. In consequence of this, the observation data vector z is also a realization of a random vector $\zeta = A\xi + \eta \in R^m$. The Bayesian model applies if, for the determination of the current vector x from z, some information about the distribution of $\xi$ can be exploited. Randomizing the vector x seems to be feasible whenever x is a representative of a vector family such that each member of this family represents n values of physical quantities at a fixed time or in a fixed location. That means, $\xi$ may for instance express the changes of x in space or in time.

However, the Bayesian approach to vector reconstruction (identification aspect) and to functional reconstruction forces us to accept the Bayesian risks

$$E\|\hat{x}(z)-x\|^2 = \int_{R^{n+m}} \|\hat{x}(z)-x\|^2 \cdot p_{\xi,\eta}(x,y)\cdot d\binom{x}{y} \tag{5.79}$$

and

$$E\,(\widehat{1^T x}(z)-1^T x)^2 = \int_{R^{n+m}} (\widehat{1^T x}(z)-1^T x)^2 \cdot p_{\xi,\eta}(x,y)\cdot d\binom{x}{y}\ , \tag{5.80}$$

which measure the mean square errors over all possible perturbations y and solution vectors x with respect to the distribution of the random vector $\binom{\xi}{\eta} \in R^{n+m}$ expressed by its density functional $p_{\xi,\eta}(x,y)$. In order to get small values for (5.79), the estimator $\hat{x}(z)$ has to reconstruct frequently occurring solution vectors x quite well, whereas the reconstruction of rare vectors x may fail.

<u>Assumption 5.21:</u>
In this Section 5.2.1., we consider the Bayesian model (5.78). It is assumed that the random vectors $\xi$ and $\eta$ are independent, i.e., $p_{\xi,\eta}(x,y) = p_\xi(x)\cdot p_\eta(y)$ for the density functionals $p_\xi(x)$, $p_\eta(y)$ and $p_{\xi,\eta}(x,y)$ of $\xi \in R^n$, $\eta \in R^m$ and $\binom{\xi}{\eta} \in R^{n+m}$, respectively. By using this assumption we suppose that the error of observation is not systematically influenced by the vector x. Moreover, let be known both $p_\xi(x)$ and $p_\eta(y)$ and hence the law of distribution of $\xi$ and $\eta$ at all. In order to designate the first and second moments, we shall apply formula (5.2) for $\eta$ and the formula

$$E\,x = \bar{x}\ , \quad \mathrm{cov}(x) = E(x-\bar{x})(x-\bar{x})^T = B \tag{5.81}$$

for $\xi$, where the covariance matrix B is supposed to be positive definite. o

**Definition 5.22:**

Estimators designated by the symbols $\hat{x}_{bay}(z)$ and $\widehat{(1^T x)}_{bay}(z)$ are called the Bayesian estimators if they minimize the Bayesian risks (5.79) and (5.80), respectively, in the class of all estimators $\hat{x}(z)$ and $\widehat{1^T x}(z)$ for the vector and functional reconstruction problem according to (5.78). ○

**Theorem 5.23:**

The Bayesian estimators are uniquely determined in the form of a conditional expectation vector

$$\hat{x}_{bay}(z)=E(x|\zeta=z):=\int_{R^n} x\cdot p(x|\zeta=z)\cdot dx=\frac{\int_{R^n} x\cdot p_\xi(x)\cdot p_\eta(z-Ax)\cdot dx}{\int_{R^n} p_\xi(x)\cdot p_\eta(z-Ax)\cdot dx} \tag{5.82}$$

and a conditional expected value

$$\widehat{(1^T x)}_{bay}(z)=E(1^T x|\zeta=z):=\frac{\int_{R^n} (1^T x)\cdot p_\xi(x)\cdot p_\eta(z-Ax)\cdot dx}{\int_{R^n} p_\xi(x)\cdot p_\eta(z-Ax)\cdot dx}, \tag{5.83}$$

where $p(x|\zeta=z)$ denotes the conditional density of $\xi$ if $\zeta$ is realized by the fixed vector z. ○

Proof: RAO [362, p.219], also [201], [138, p. 42].

Bayesian estimators are in almost all cases nonlinear even is A is linear. The numerical evaluation of formulae (5.82) and (5.83) needs computing multiple integrals. Problems arise from the fact that the region of integration is unlimited. Therefore, the nonlinear estimators of Theorem 5.23 are of minor practical importance. Frequently, it seems to be satisfactory to confine the consideration to well-known distribution families (e.g. normal distribution).

Now we are going to study the linear Bayesian model

$$A\,x + y = z\,,\quad A\in R^{m\times n},\ x\in R^n,\ y,z\in R^m \tag{5.84}$$

as a variant of (5.78) with linear operator A. Models of this kind have also been discussed by DAVIES [83-84], MURAVEVA [325], RAO [361], STRAND/WESTWATER [414-15], SWAMY [420] and TURCHIN et al. [448]. If both the random vectors $\xi$ and $\eta$ in the linear Bayesian model possess a normal (Gaussian) distrubution, i.e.,

$$\xi\sim\mathcal{N}(\bar{x},B)\ :\ p_\xi(x)\propto\exp(-\tfrac{1}{2}(x-\bar{x})^T B^{-1}(x-\bar{x})) \tag{5.85}$$

and

$$\eta\sim\mathcal{N}(0,C)\ :\ p_\eta(y)\propto\exp(-\tfrac{1}{2}y^T C^{-1}y)\,, \tag{5.86}$$

then Bayesian estimators become linear ones. In this special case, one can simplifiy the expressions of $\hat{x}_{bay}(z)$ and $\widehat{(1^T x)}_{bay}(z)$.

Theorem 5.24:

Consider the Bayesian linear model (5.84) with Gaussian random vectors $\zeta$ and $\eta$ satisfying (5.85) and (5.86). Then,

$$\hat{x}_{bay}(z) = \hat{x}_{sr}(z) := \bar{x} + V_{sr}(z - A\bar{x}), \quad V_{sr} := BA^T(ABA^T + C)^{-1} = (A^T C^{-1} A + B^{-1})^{-1} A^T C^{-1}, \tag{5.87}$$

with the resulting Bayesian risk

$$E\|\hat{x}_{bay}(z) - x\|^2 = \operatorname{trace}(B) - \operatorname{trace}(BA^T(ABA^T + C)^{-1}AB) = \operatorname{trace}((A^T C^{-1} A + B^{-1})^{-1}), \tag{5.88}$$

and

$$(\widehat{1^T x})_{bay}(z) = 1^T \cdot \hat{x}_{sr}(z), \tag{5.89}$$

with the associated Bayesian risk

$$E(1^T \hat{x}_{sr}(z) - 1^T x)^2 = (B1, 1) - ((ABA^T + C)^{-1} AB1, AB1) = ((A^T C^{-1} A + B^{-1})^{-1} 1, 1). \quad \circ \tag{5.90}$$

Proof:

We acquiesce in proving the formulae (5.87) and (5.88) here. After substituting $u := x - \bar{x}$ we have $p(x|\zeta = z) \propto \exp(-\frac{1}{2}(u^T B^{-1} u + (z - A\bar{x} - Au)^T C^{-1}(z - A\bar{x} - Au)))$. The exponent can be rewritten as follows: $u^T B^{-1} u + (z - A\bar{x} - Au)^T C^{-1}(z - A\bar{x} - Au) = u^T(A^T C^{-1} A + B^{-1})u + (z - A\bar{x})^T C^{-1}(z - A\bar{x}) - 2u^T A^T C^{-1}(z - A\bar{x}) = (u - V_{sr}(z - A\bar{x}))^T \cdot (A^T C^{-1} A + B^{-1})(u - V_{sr}(z - A\bar{x})) + (z - A\bar{x})^T C^{-1}(z - A\bar{x}) - (V_{sr}(z - A\bar{x}))^T \cdot (A^T C^{-1} A + B^{-1})(V_{sr}(z - A\bar{x}))$. As the last obtained formulation shows, $p(x|\zeta = z)$ is symmetric with respect to $\tilde{u} := u - V_{sr}(z - A\bar{x})$ at $\tilde{u} = 0$. Therefore, $\int_{R^n} \tilde{u} \cdot p(x|\zeta = z) \cdot dx = 0$ and $E(x|\zeta = z) = \int_{R^n} x \cdot p(x|\zeta = z) \cdot dx = \int_{R^n} (\tilde{u} + \bar{x} + V_{sr}(z - A\bar{x})) \cdot p(x|\zeta = z) \cdot dx = \int_{R^n} (\bar{x} + V_{sr}(z - A\bar{x})) \cdot p(x|\zeta = z) \cdot dx = \bar{x} + V_{sr}(z - A\bar{x})$ due to $\int_{R^n} p(x|\zeta = z) \cdot dx = 1$. The formula (5.88) is a consequence of (5.17), where we substitute $V := V_{sr}$, $v := \bar{x} - V_{sr}A\bar{x}$. Namely,
$E\|\hat{x}_{sr}(z) - x\|^2 = E\|(V_{sr}A - I)(x - \bar{x})\|^2 + \operatorname{trace}(V_{sr} C V_{sr}^T) =$
$= \operatorname{trace}\{(V_{sr}A - I)B(V_{sr}A - I)^T + V_{sr} C V_{sr}^T\} = \operatorname{trace}((A^T C^{-1} A + B^{-1})^{-1}) =$
$= \operatorname{trace}(B) - \operatorname{trace}(BA^T(ABA^T + C)^{-1}AB)$ (cf. formula (5.40)). #

Remark 5.25:

The estimation of x based on $\hat{x}_{sr}(z)$ is called the statistical regularization, since the Bayesian estimator according to a normal linear Bayesian model can be considered to be a generalization to the regularization procedure (4.80). Note that $\hat{x}_{sr}(z)$ minimizes the functional

$$F_{sr}(x) := (C^{-1}(z-Ax), z-Ax) + (B^{-1}(x-\bar{x}), x-\bar{x}) \tag{5.91}$$

in $R^n$. This indicates that the moment information (5.85) suffices to determine the term $\alpha \cdot \Omega(x)$ of Tikhonov's stabilizing functional (4.1) with $\Omega$ according to formula (4.64). Provided that $C = \sigma^2 I$, the mean vector $\bar{x}$ and the covariance matrix inverse $\Phi := B^{-1}$ form the evaluation functional. Moreover, $\alpha := \sigma^2$ embodies the adapted regularization parameter (see formula (5.91)). Statistical regularization is a version of generalized ridge estimation (see (5.44)) whenever $\bar{x}=0$. o

We shall continue to consider the linear Bayesian model (5.84), where the coming remarks are concentrated on linear estimators (5.3) and (5.4). Here, Gaussian distribution (5.85) and (5.86) is not required in the sequel. However, we are especially interested in the Bayesian generalized unbiased estimators defined below.

Definition 5.26:
An estimator $\hat{x}(z)$ according to the Bayesian model is said to be a Bayesian generalized unbiased estimator in the vector reconstruction case if, independently of the present mean vector $\bar{x} \in R^n$,

$$E\, \hat{x}(z) = \bar{x}\ . \tag{5.92}$$

In the functional reconstruction case, an estimator $\widehat{l^T x}(z)$ is termed a Bayesian generalized unbiased estimator if

$$E(\widehat{l^T x}(z)) = l^T \bar{x} \tag{5.93}$$

holds. o

Linear estimators (5.3) and (5.4) satisfy (5.92) and (5.93) iff, respectively, $(VA-I)\bar{x}+v = 0$ and $(w^T A - l^T)\bar{x} + \varrho = 0$. We can easily derive that, for the Bayesian risk of linear estimators,

$$E\|\hat{x}(z)-x\|^2 = \mathrm{trace}\{(VA-I)B(VA-I)^T + VCV^T\} + \|(VA-I)\bar{x}+v\|^2 \tag{5.94}$$

and

$$E((\widehat{l^T x}(z)) - l^T x)^2 = (B(A^T w - l), A^T w - l) + (Cw, w) + ((A^T w - l, \bar{x}) + \varrho)^2. \tag{5.95}$$

If and only if the estimators are Bayesian generalized ones, the third terms of the right hand sides in Eqs. (5.94) and (5.95) vanish.

Definition 5.27:
Estimators $\hat{x}_{lbay}(z)$ and $(\widehat{l^T x})_{lbay}(z)$ according to the linear Bayesian model (5.84) are named the linear Bayesian estimators if they minimize the Bayesian risks (5.79) and (5.80) in the class of linear estimators (5.3) and (5.4), respectively. o

By exploiting formula (5.94) together with Lemma 5.19 ($G := I$), in the vector case, and formula (5.95) together with Lemma 4.32, in the functional case, we are able to verify the structure of linear Bayesian estimators.

Theorem 5.28:

The linear Bayesian estimators are uniquely determined. They coincide with Bayesian estimators in the case of Gaussian distributed random vectors $\xi$ and $\eta$. That is, for a Gaussian model holds:

$$\hat{x}_{lbay}(z) = \hat{x}_{sr}(z) \tag{5.96}$$

and

$$(\widehat{1^T x})_{lbay}(z) = 1^T \cdot \hat{x}_{sr}(z) \tag{5.97}$$

(cf. Theorem 5.24). o

Remark 5.29:

It is characteristic of linear estimation in linear Bayesian models that, independently of the distribution type of $\xi$ and $\eta$, statistical regularization estimators minimize the Bayesian risks. The only requirement on the occurring random vectors is concerned with means and covariance matrices regarding the formulae (5.2) and (5.81). If $\xi$ and $\eta$ are non-Gaussian random vectors, the Bayesian risk may further be reduced by using appropriate nonlinear estimators (see Theorem 5.23). Unfortunately, the required behaviour of $p_\xi(.)$ and $p_\eta(.)$ is rarely known in detail for tasks arising from application. o

At the end of this Sec. 5.2.1., we still place at our disposal two lemmas which are needed in the next section.

Lemma 5.30:

Let $\bar{x}=0$ and $M\in R^{n\times n}$ a positive semidefinite symmetric matrix. Then, for the ridge-like estimator

$$\hat{x}_M(z) := V_M z\ , \quad V_M := MA^T(AMA^T+C)^{-1}\ , \tag{5.98}$$

the Bayesian risk is limited uniformly with respect to M:

$$E\|\hat{x}_M(z)-x\|^2 \leq \operatorname{trace}(B+(A^TC^{-1}A)^+)\ . \quad o \tag{5.99}$$

Proof:

Firstly, we prove the inequality (5.99) for positive definite matrices M. Thus, $\hat{x}_M(z)=(A^TC^{-1}A+M^{-1})^{-1}A^TC^{-1}z$ (see (5.44)) and $E\|\hat{x}_M(z)-x\|^2 = \operatorname{trace}\{(A^TC^{-1}A+M^{-1})^{-1}(M^{-1}BM^{-1}+A^TC^{-1}A)(A^TC^{-1}A+M^{-1})^{-1}\}$ $= \operatorname{trace}\{B^{1/2}(M^{1/2}A^TC^{-1}AM^{1/2}+I)^{-2}B^{1/2}\} + \operatorname{trace}\{((A^TC^{-1}A)^+)^{1/2}A^TC^{-1}A\cdot$ $\cdot(A^TC^{-1}A+M^{-1})^{-2}A^TC^{-1}A((A^TC^{-1}A)^+)^{1/2}\}$. If $M_1\in R^{n\times n}$, $M_2\in R^{n\times n}$ are arbitrary quadratic matrices and $M_3\in R^{n\times n}$ is a positive semidefinite symmetric matrix, then we have $\operatorname{trace}(M_3M_1)\leq \operatorname{trace}(M_3M_2)$ whenever

$(M_1x,x) \leq (M_2x,x)$ for all $x \in R^n$ (see [325]). This proves the inequality (5.99) if we recognize $((M^{1/2}A^TC^{-1}AM^{1/2}+I)^{-2}x,x) \leq (x,x)$, $((A^TC^{-1}A+M^{-1})^{-2}x,x) \leq (((A^TC^{-1}A)^+)^2x,x)$ and hence $(A^TC^{-1}A(A^TC^{-1}A+M^{-1})^{-2}A^TC^{-1}Ax,x) \leq (x,x)$ for all $x \in R^n$. Since any positive semidefinite matrix is a limit of a sequence of positive definite matrices, the continuous dependence of $E\|\hat{x}_M(z)-x\|^2$ upon M (see Lemma 5.31) yields the validity of (5.99) in the general case. #

Lemma 5.31:

Under the assumptions of Lemma 5.30 we can state the Lipschitz conditions

$$\|V_{M+\tilde{M}} - V_M\| \leq c_1(A,C,M)\cdot\|\tilde{M}\| \tag{5.100}$$

and

$$|E\|\hat{x}_{M+\tilde{M}}(z)-x\|^2 - E\|\hat{x}_M(z)-x\|^2| \leq c_2(A,B,C,M)\cdot\|\tilde{M}\|, \tag{5.101}$$

where $M \in R^{n\times n}$ and $M+\tilde{M} \in R^{n\times n}$ are supposed to be symmetric and positive definite. The values $c_1 < \infty$ and $c_2 < \infty$ do not depend on $\tilde{M}$. o

Proof: [201, p.30].

## 5.2.2. Empirical Statistical Regularization

This section is devoted to the empirical Bayesian noisy data problem (see Definition 2.65) and to empirical modifications to statistical regularization. We examine the case that the mean vector $\bar{x}$ and the covariance matrix B are not known a priori. Instead a sample of realizations of $\xi$ is submitted, based on which one can estimate the unknown moments. When applying the statistical regularization estimator with empirical moments, we have to take into consideration an additional error. Here, we shall outline only some basic relations concerning the use of empirical data for statistical regularization. For more detailed and substantially extended studies of empirical statistical regularization, the reader is referred to other papers of the author (see [200-02], [208], [210] and [138, Sec.1.4.]).

Assumption 5.32:

In this Sec. 5.2.2., we consider the linear Bayesian model (5.84) subject to (5.2) and (5.81). Suppose that $C \in R^{m\times m}$ is known, whereas $\bar{x}$ and B are a priori unknown. The missing $n+\frac{n(n+1)}{2}$ real numbers determining the first and second distribution moments of $\xi$ are to be estimated from a given sample of direct observations $\{x^{(i)}\}_{i=1}^k$ with a sample size k. Each vector $x^{(i)} \in R^n$ is a realization of the random vector $\xi^{(i)}$. For all $i=1(1)k$, the random vectors $\xi^{(i)}$ and $\xi$ are identically distributed. Moreover, we assume the vectors $\xi, \xi^{(1)}, \dots, \xi^{(k)}$ and $\eta$ to be pairwise independent. o

We are going to examine empirical linear Bayesian estimators, i.e., we apply the empirical method (see H. and O. BUNKE [45]) to linear Bayesian estimators for the model under consideration. In order to do this, the statistical regularization $\hat{x}_{sr}(z) = \bar{x}+V_{sr}(z-A\bar{x})$ (see formula (5.87)) that represents the linear Bayesian estimator is replaced by an available neighbouring estimator

$$\hat{x}_{esr}(z) := \hat{\bar{x}}+V_{esr}(z-A\hat{\bar{x}}) \ , \quad V_{esr} := \hat{B}A^T(A\hat{B}A^T+C)^{-1} \quad . \tag{5.102}$$

The vector $\hat{\bar{x}}\in R^n$ and the matrix $\hat{B}\in R^{n\times n}$ are understood to be well-behaved estimates of $\bar{x}$ and B derived from the sample $\{x^{(i)}\}_{i=1}^{k}$.

Remark 5.33:

An empirical modification to statistical regularization of the form (5.102) is termed the empirical statistical regularization. In this section, estimators of such kind are studied for linear models of full rank and of deficient rank of A in a unified manner (see Sec. 5.3.). Mathematical statistics is preferably concerned with the full-rank case (see e.g. BUNKE/GLADITZ [46], SWAMY [420], RAO [363]). We will only mention results for the vector identification case (see also [201],[208],[210]). A functional identification variant can be derived easily. o

Frequently, the so-called empirical moments

$$\bar{x}_{emp} := \frac{1}{k} \sum_{i=1}^{k} x^{(i)} \tag{5.103}$$

and

$$B_{emp} := \frac{1}{k-1} \sum_{i=1}^{k} (x^{(i)}-\bar{x}_{emp})(x^{(i)}-\bar{x}_{emp})^T \tag{5.104}$$

serve as estimates of mean vector $\bar{x}$ and covariance matrix B.

Lemma 5.34:

The empirical moments (5.103) and (5.104) are unbiased and consistent estimates of $\bar{x}$ and B, respectively, i.e.,

$$E\,\bar{x}_{emp} = \bar{x} \quad , \quad E\,B_{emp} = B \tag{5.105}$$

and

$$\text{p-lim}_{k\to\infty}\ \bar{x}_{emp} = \bar{x} \ , \quad \text{p-lim}_{k\to\infty}\ B_{emp} = B \ , \tag{5.106}$$

whenever Assumption 5.32 holds. o

As to the proof of Lemma 5.34 and other features of empirical moments, see e.g. [201] or [241]. Note that the symbol $\text{p-lim}_{k\to\infty}$ denotes the stochastic convergence of a sequence of random elements.

**Definition 5.35:**

A sequence of random elements (random variables, vectors or matrices) $\xi^{(k)}$ with realizations $x^{(k)}$ is stochastically convergent to the random element $\xi$ with realizations x if

$$\lim_{k\to\infty} P(\|x^{(k)}-x\|\geq\varepsilon) = 0 \text{ , for all } \varepsilon>0. \quad \circ \tag{5.107}$$

In the paper [46], BUNKE and GLADITZ stated that the symmetric matrix $B_{emp}$ may possess negative eigenvalues for which no statistical interpretation exists. Both authors proposed the use of a projection matrix

$$\tilde{B}_{emp} : \|\tilde{B}_{emp} - B_{emp}\|_F = \min_{\tilde{\tilde{B}}\in R^{n\times n}_{psd}} \|\tilde{\tilde{B}} - B_{emp}\|_F \tag{5.108}$$

instead of $B_{emp}$ itself. The Frobenius norm projection $\tilde{B}_{emp}$ of $B_{emp}$ into the set $R^{n\times n}_{psd}$ of positive semidefinite quadratic symmetric matrices of dimension n is uniquely determined. If $B_{emp}= \sum_{i=1}^{n}\lambda_i\cdot u^{(i)}\cdot(u^{(i)})^T$ with $\lambda_i$ and $u^{(i)}$, i=1(1)n, designating the eigenvalues and associated orthonormalized eigenvectors of the matrix $B_{emp}$, then $\tilde{B}_{emp} = \sum_{\lambda_i\geq 0} \lambda_i\cdot u^{(i)}\cdot(u^{(i)})^T$ . Owing to the convexity of the set $R^{n\times n}_{psd}$ , we have $\|\tilde{B}_{emp} - B\|_F \leq \|B_{emp} - B\|_F$ and thus $\text{p-lim}_{k\to\infty} \tilde{B}_{emp} = B$ . On the other hand, the estimate $\tilde{B}_{emp}$ defined by formula (5.108) is consistent with respect to B. However, the unbiasedness of $B_{emp}$ gets lost if we substitute $\tilde{B}_{emp}$ for $B_{emp}$.

In the sequel, consider the empirical statistical regularization (5.102) subject to

$$\hat{\bar{x}} := \bar{x}_{emp} \text{ , } \hat{B} := \tilde{B}_{emp} \text{ .} \tag{5.109}$$

We will prove that empirical statistical regularization is asymptotically optimal, i.e., the Bayesian risk of empirical statistical regularization is stochastically convergent to the Bayesian risk of statistical regularization as the sample size k tends to infinity. So, we have

$$\text{p-lim}_{k\to\infty} \|\hat{x}_{esr}(z)-x\|^2 = E\,\|\hat{x}_{sr}(z)-x\|^2 \text{ .} \tag{5.110}$$

Moreover, even the mixed risk

$$E_{mix}\,\|\hat{x}_{esr}(z)-x\|^2 := \int_{(R^n)^k} E\|\hat{x}_{esr}(z)-x\|^2\cdot p_{\xi^{(1)}}(x^{(1)})\ldots p_{\xi^{(k)}}(x^{(k)})\cdot d\,x^{(1)}\cdot\ldots\cdot d\,x^{(k)} \text{ ,} \tag{5.111}$$

averaging the Bayesian risk over all possible samples of size k, tends to the optimal value $E\|\hat{x}_{sr}(z)-x\|^2$, i.e.,

$$\lim_{k\to\infty} E_{mix} \|\hat{x}_{esr}(z)-x\|^2 = E \|\hat{x}_{sr}(z)-x\|^2 . \tag{5.112}$$

Theorem 5.36:

If we define the empirical statistical regularization by the formulae (5.102) and (5.109), then it follows (5.110) as well as (5.112). The risks of empirical statistical regularization approximate the risk of statistical regularization arbitrarily well if the sample size increases unlimitedly. o

Proof:

We have $E\|\hat{x}_{esr}(z)-x\|^2 = \|(V_{esr}A-I)(\hat{\bar{x}}-\bar{x})\|^2 + \mathrm{trace}\{(V_{esr}A-I)\cdot B\cdot \cdot(V_{esr}A-I)^T + V_{esr}\cdot C\cdot V_{esr}^T\}$ with $\|V_{esr}A-I\|\leq 1$ (cf. the proof of Lemma 5.30). Then, $\text{p-}\lim_{k\to\infty} \|(V_{esr}A-I)(\hat{\bar{x}}-\bar{x})\|^2 = 0$ is a consequence of the consistency of the estimate $\bar{x}_{emp}$ with respect to $\bar{x}$ (see Definition 5.35). Now consider the generalized ridge estimator $\hat{x}_{gridge}(z) = \hat{B}A^T(A\hat{B}A^T+C)^{-1}z$ subject to $\hat{B} := \tilde{B}_{emp}$. The term $\mathrm{trace}\{(V_{esr}A-I)B(V_{esr}A-I)^T + V_{esr}CV_{esr}^T\}$ can be interpreted as the Bayesian risk of the above estimator $\hat{x}_{gridge}(z)$ with $\bar{x}=0$. Therefore, the Lemmas 5.30 and 5.31 apply. Due to formula (5.101) we obtain $P(|\mathrm{trace}\{(V_{esr}A-I)B(V_{esr}A-I)^T+V_{esr}CV_{esr}^T\} - \mathrm{trace}\{(V_{sr}\cdot A-I) \cdot B(V_{sr}\cdot A-I)^T + V_{sr}\cdot C\cdot V_{sr}^T\}| \geq \varepsilon) \leq P(\|\tilde{B}_{emp}-B\| \geq \frac{\varepsilon}{c_2(A,B,C)})$. Hence, the consistency of $\tilde{B}_{emp}$ with regard to B causes the condition (5.110). In order to derive (5.112) from (5.110), we can use Lebesgue's theorem. This theorem asserts, for a sequence of random variables $\xi^{(k)}$ with realizations $x^{(k)}$, that $\text{p-}\lim_{k\to\infty} \xi^{(k)} = \xi$ implies $\lim_{k\to\infty} E\, x^{(k)} = x$ whenever there is a majorant $c>0$ with $P(|x^{(k)}| \leq c) = 1$ and $k=1,2,\ldots$ . Consequently, (5.112) becomes clear if we recall Lemma 5.30. Formula (5.99) yields a majorant for the second term of $E\|\hat{x}_{esr}(z)-x\|^2$ (see the first formula of this proof). Thus, we have $E_{mix}\, \mathrm{trace}\{(V_{esr}A-I)B(V_{esr}A-I)^T + V_{esr}CV_{esr}^T\} = \mathrm{trace}\{(V_{sr}A-I)B(V_{sr}A-I)^T+V_{sr}CV_{sr}^T\}$ . Moreover,

$$E_{mix} \|(V_{esr}A-I)(\hat{\bar{x}}-\bar{x})\|^2 \leq E_{mix} \|\hat{\bar{x}}-\bar{x}\|^2 = E_{mix}\, \mathrm{trace}\{(\hat{\bar{x}}-\bar{x})(\hat{\bar{x}}-\bar{x})^T\} = \frac{1}{k}\cdot\mathrm{trace}(B) \longrightarrow 0 \text{ as } k\to\infty .$$

Namely, $\hat{\bar{x}}-\bar{x}$ is a centralized random vector with covariance matrix $\frac{1}{k}\cdot B$ . This completes the proof. #

**Remark 5.37:**

It is very difficult to express the influence of estimated moments to the total error of estimation if a finite number k of sample vectors is given. Monte Carlo simulation experiments can help to evaluate the mixed risk (5.112) depending on k. For the applied problem discussed in Sec. 5.3., a computer simulation based on Gaussian pseudorandom numbers was performed by the author (see [210]). The results pointed out that the rate of convergence concerning the limiting process of formula (5.112) is almost linear:

$$E_{mix} \|\hat{x}_{esr}(z)-x\|^2 \approx E \|\hat{x}_{sr}(z)-x\|^2 + \frac{\varkappa(A,B,C)}{k} \quad . \tag{5.113}$$

It seems to be clear that the influence of mean estimation to the whole estimation error is small compared to the covariance estimation, since in the former case (mean estimation) n unknowns are to be determined, whereas $\frac{n(n+1)}{2}$ free values correspond to the latter case (covariance estimation). o

The evaluation of empirical Bayesian estimators by a posteriori risks $E(\|\hat{x}_{esr}(z)-x\|^2 | \xi^{(1)}=x^{(1)},\ldots,\xi^{(k)}=x^{(k)})$ may provide further interesting propositions (see [200-02], [138]). However, a noninformative (see KLIMOV [241]) or informative (see EVANS [120], LINDLEY/SMITH [273]) a priori distribution of $\bar{x}$ and B has to be assumed. We avoid presenting the mathematical calculus of multivariate Gaussian, Student and Wishart distributions required for those studies in order not to burden the text inappropriately.

## 5.3. On an Example Concerning the Stochastic Modelling

We are concerned with mathematical problems regarding the remote sounding of vertical atmospheric temperature profiles by satellite data. In 1970, KONDRATEV and TIMOFEEV published their book [250] on thermal sounding that also involves a numerical and statistical approach to the estimation of temperature profiles from noisy observations of the intensity of thermal radiation with respect to different wavelengthes. If $\varkappa=\langle x(t),\ \underline{t}\leq t\leq\bar{t}\rangle$, with altitude t, represents the unknown vertical temperature profile at a given time and $\mathscr{b}=\langle b(s),\ \underline{s}\leq s\leq\bar{s}\rangle$ the simultaneously measured intensity of thermal radiation with wavelength s caused by that profile, then a linear Fredholm integral equation of the first kind (2.16) with a smooth kernel describes the associated inverse problem $\mathscr{A}\varkappa=\mathscr{b}$ . The positive kernel function k(s,t) is assigned to the transmissivity of an atmospheric layer possessing the altitude t with respect to thermal radiation of wavelength s.

For the construction of an appropriate discretization model, we have to consider that the formulae (2.33) and (2.37) are used for the experimental design operator $Q^m$ and for the a priori discretization operator $P^n$. The vector x to be reconstructed includes weighted average temperatures $x_i := \int^{t_{i+1}}_{t_i} f_1(|t-t_i'|)\cdot x(t)\cdot dt$, $i=1(1)n$ of vertical subprofiles.
On the other hand, the observed data $b_j := \int^{s_{j+1}}_{s_j} f_2(|s-s_j'|)\cdot b(s)\cdot ds$ characterize, for $j=1(1)m$, mean values of the radiation intensity over some wavelength intervals. The monotonically nonincreasing weighting functions $f_1(.)$ and $f_2(.)$ are of different nature. The function $f_1$ may be prescribed, whereas $f_2$ depends on the used measuring tool. Here, the numbers $t_i'$ and $s_j'$ designate the mid-points of the subintervals $[t_i, t_{i+1}]$ and $[s_j, s_{j+1}]$, respectively. In order to get a matrix $A \in R^{m \times n}$, the corresponding entries $a_{ji}$ averaging over the rectangle $[t_i, t_{i+1}] \times [s_j, s_{j+1}]$ with respect to the transmissivity may be won from a table of atmospheric properties. It seems to be reasonable to assume that A embodies the expectation of a random matrix such that the discretization error $Q^m \mathcal{A} \varkappa - AP^n \varkappa$ approximately has the behaviour of a centralized random vector. Then, the linear model (5.1) applies if the satellite data involving inevitable errors of observation in each component are gathered into the vector z.

Due to its physical background the thermal sounding problem is an identification one. Both the vector reconstruction problem and the functional reconstruction problem are of interest for meteorologists. A simultaneous identification of the entire temperature profile from satellite data can help to improve weather forecasting. On the other hand, functional estimates may also provide knowledge of local atmospheric properties (e.g. mean temperatures of troposphere or tropopause). Note that at least in the 1970's the measuring tools on board of satellites usually permitted a small number m of well-seperated observations only. Hence, estimators for the deficient-rank case $m<n$ may be of great importance. The author's thesis [201] is predominantly devoted to the deficient-rank estimation in satellite data problems.

Time dependent randomization of temperature profiles related to a fixed area of a continent is promising whenever there are chances of obtaining mean $\bar{x}$ and covariance matrix B of the associated random vector. Thus, the linear Bayesian model (5.84) becomes active. If a meteorological base works in this area and has taken measurements for a long time, a sample of repeated temperature observations is available. The direct measurements have been obtained by expensive baloon ascents. Thus, the empirical statistical regularization seems to be applicable and frequently optimal in the class of practicable estimation procedures for the present remote sounding problem. For numerical results that illustrate the different estimators and their accuracy by means of constellations concerning the temperature reconstruction, the reader is referred to [200-02] and [210]. Generally speaking, there are good chances of recovering the vertical temperature profiles with a standard deviation of about 1...2 degrees in any component.

# 6. A Unified Numerical Approach to Nonlinear Inverse Problems

## 6.1. Classifying Inverse Problems with Respect to the Amount of Computation

In this chapter, we delve more deeply into those aspects of the regularized solution of discretized inverse problems that are concerned with the required computational work. Let us begin by classifying the different classes of problems with respect to the amount of computation. From Chapter 4 we have learned that identification and control problems are regularized in a unified manner. Consequently, for our classification, we only have to consider the space dimensions m and n of the discretized inverse problem and intrinsic features of the operator A and of the domain D. Numerical experience suggests a classification as follows:

Figure 6.0:

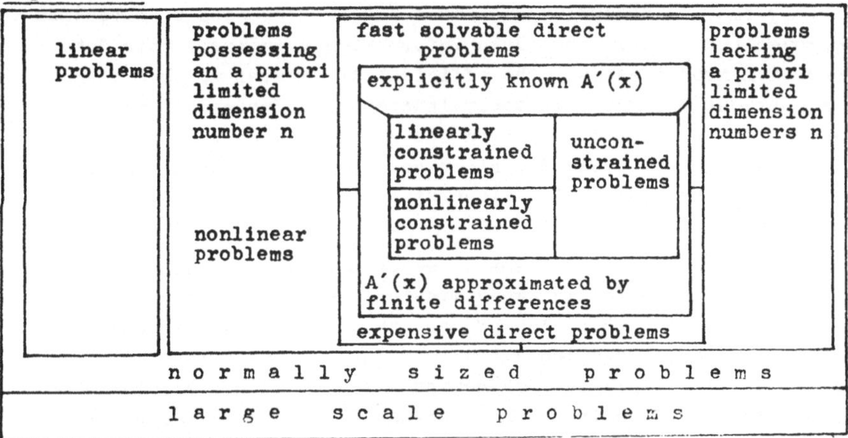


The total amount of work for the approximate solution of a discretized inverse problem depends on the answers to the following questions:

Q1: Ist the problem normally sized, i.e., are the chosen dimension numbers m and n small enough in order to perform all the required computations without a substantial external data transfer ?

Q2: Is the operator A and therefore the discretized inverse problem linear or nonlinear ?

Q3: Are there arguments concerning the practical background of the problem, which allow us to confine the a priori discretization to a small number n of unknowns ?

Q4: How expensive is the computation of Ax for given x (costs of solving the direct problem) ?

Q5: Are the Fréchet derivatives A′(x) explicitly known or are they to be approximated by finite differences of Ax ?

Q6: Is the problem unconstrained, linearly constrained or nonlinearly constrained ?

The majority of inverse problems can be discretized so that normally sized problems arise. It is no use choosing the number n very large if a function of one variable is discretized by an n-dimensional skeleton vector. Due to ill-posedness and ill-conditioning effects a fine grid in a priori discretization fails to imply an improved accuracy of approximate solutions (see Sec.2.2.1). If the submitted data vector z has very large dimension m, then one should check whether there is a modified data vector of considerably smaller dimension such that the problem condition becomes improved but no principal data information gets lost. This data reduction method seems to be efficient if several neighbouring observations are not completely independent. Instead of measurements at many times of a period one should use an average value. From the point of view of statistics, the average value is even expected to have a smaller dispersion than any of the single observation values.

However, large scale problems arise of necessity if the geometry of the unknown object is two-dimensional or three-dimensional (e.g. for the determination of density distributions in computerized tomography - see Sec.2.2.3.- or for the determination of spatially varying geological phenomena - see [C12]). Then, the numbers m and n can be in the many thousands, although the a priori discretization grid is rather coarse. In fact, there exists no sharp boundary between normally sized problems and large scale problems. The used computer hardware and software is of great importance in this context. With the accelerated developemnt of computers the problem sizes, which are small enough as to be completely handled by a personal computer, will continuously grow. But from the point of view of data error influence and ill-posedness it is always recommended to exploit any chance to decompose large inverse problems into a set of smaller problems.

Linear discretized inverse problems are not difficult to solve numerically by regularization methods. The numerical linear algebra is an advanced subbranch of numerical mathematics. On the other hand, the treatment of nonlinear inverse problems requires the repeated numerical solution of nonlinear and frequently nonconvex optimization problems. Difficulties arising from the computational evaluation of approximate regularized solutions are raised to a higher power if we leave the linear inverse problems and turn to nonlinear ones. Linear integral equations of the first kind and their discrete counterparts are considered in Secs. 2.2.1. and 2.2.2.. For nonlinear examples, we refer back to the Secs. 2.2.4. through 2.2.8..

Regularization routines for solving nonlinear discretized inverse problems are based on the repeated iterative numerical solution of nonlinear optimization problems (see the Algorithms 4.17 and 4.26). The whole computational process of regularization requires the many times solution of the associated direct problem, i.e., the recomputation of Ax for various different vectors $x \in D$. Therefore, it is of great importance whether it takes us a millisecond, a second, a minute or an hour to solve one direct problem. Fast solvable direct problems, for example, correspond to explicitly given elementary procedures for computing Ax (see formula (2.52) of Sec.2.2.4. or formula (2.82) of Sec.2.2.6.). In such a case, large scale computers perform hundreds of direct problem solutions during one second. On the other hand, if the nonlinear operator A is implicitly given by an initial-boundary value problem associated with partial differential equations (see Secs. 2.2.5. and 2.2.7.) or by an eigenvalue problem (see Sec. 2.2.8.), then the solution of one direct problem is expensive. In dependence of the special choice of dimension numbers, used grids and direct solvers, the computation time, for one vector Ax, may considerably exceed one second.

Frequently, fast solvable direct problems have the additional advantage that the Fréchet derivative $A'(x)$ is also explicitly known and fast computable (see Theorem 2.55 and formula (2.83)). This helps compute gradients $\nabla_x F(x,\alpha)$ (see formulae (4.1), (4.66)), which are needed for the iterative numerical solution of problem (4.2). Exceptionally, expensive direct problems may also correspond to explicitly available derivatives. Thus, the inverse eigenvalue problem of Sec. 2.2.8. yields $A'(x)$ in a calculable form (see formula (2.93)). Whenever $A'(x)$ fails to be expressed explicitly, derivative-free methods (see SCHWETLICK [394, Chap.9]) should be applied. So, $A'(x)$ is to be approximated by finite differences of vectors Ax. Unfortunately, we have to solve n+1 direct problems in order to get one Fréchet derivative. Thus, the computational process is very expensive, in particular if any direct problem solution requires much work. However, there are some tricks aimed at reducing the total amount of work. A first trick is related to the class of inverse problems in partial differential equations. From the solution of the so-called adjoint problem (see [156] or [193]) to the direct initial-boundary value problem under consideration one can derive the needed gradient. A second trick is based on a cost reduction by using a combination of fine and coarse grids. This technique will be presented in Sec. 6.4..

The numerical treatment of nonlinear discretized inverse problems with expensive corresponding direct problems may touch the usually given limitations of admissible computational cost. If, for example, n=20 unknowns are to be determined for the skeleton of a material function and it takes us 30 seconds to solve the associated direct problem, then by using a derivative-free minimization routine the computer is occupied more than one hour in performing one iteration step. Therefore, it is deriable to have an a priori limitation of n to a small number of unknowns, which is justified by physically or technically motivated background information (see e.g. formulae (2.76) or (2.77) for the determination of heat conductivity or heat transfer functions).

Finally, the character of D also influences the expense of computing regularized solutions. Unconstrained problems allow the application of unconstrained minimization techniques, whereas constraints of any form require an additional amount of work (see Sec. 6.2.). Linear constraints (lower and upper bounds, monotonicity and convexity requirements on the function to be determined) are fairly easy to handle (see [441]). On the other hand, nonlinear constraints cause much more difficulties, but they seldom occur in the modelling of applied inverse problems. In the coming section, we are going to propose a general strategy for the numerical computation of regularized solutions to nonlinear discretized inverse problems. This strategy is independent of the linearity or nonlinearity of arising constraints. It is evident that in evaluating regularized solutions there are better numerical methods for any special case. But the study of Sec. 6.2. helps formulate a software strategy in Sec. 6.3., which is aimed at the solution of very different discretized inverse problems in a unified manner.

## 6.2. A Numerical Route for Computing Regularized Solutions

In this section, we refer back to the algorithms 4.17 and 4.26. From the point of view of numerical mathematics a procedure that helps to calculate the regularized solution $x_{(\alpha)}$, for given $\alpha$, (see Algorithm 4.17, Step1) and a procedure for evaluating $x_{[\lambda]}$ (see Algorithm 4.26, Step1) are still lacking. We are going to pay attention to the single-parameter case $x_{(\alpha)}$ and propose a strategy for the numerical solution of the optimization problem (4.2). subject to (4.1) and (4.64):

$$\underset{x \in D}{\text{minimize}} \left\{ \|Ax-z\|^2 + \alpha(\Phi(x-\bar{x}), x-\bar{x}) \right\}. \tag{6.1}$$

As we will see, the derived method can also immediately be applied to the multi-parameter problem

$$\underset{x \in D}{\text{minimize}} \left\{ \sum_{i=1}^{k} \lambda_i \|A_i x - z^{(i)}\|^2 + (\Phi(x-\bar{x}), x-\bar{x}) \right\}. \tag{6.2}$$

This is due to the fact that our method postulates a sum of square terms, i.e., a Euclidean norm square, in order to characterize the structure of the objective functional which is to be minimized.

We have some good arguments that suggest transforming the constrained optimization problem (6.1) into an unconstrained one by using the method of exterior quadratic penalty functions.

- Penalty function methods may be applied to both linear and nonlinear constraints. Moreover, posterior estimations of the precision of penalty method solutions are available (see GROSSMANN/KAPLAN [182, Chap.3]).
- In real life problems, the moderate accuracy of measurement data and the empirical character of a priori information generating D and $\Omega$ seldom require the use of high penelty levels (large values of the penalty parameter). However, ill-conditioned unconstrained optimization problems arise from the penalty function method if and only if high penalty levels are active. Note that an unconstrained optimization problem is called ill-conditioned at a minimum point if the Hessian of the objective functional is ill-conditioned at this point (as for numerical difficulties resulting from the ill-conditioning of optimization problems, see [182, Chap.8]).
- If exceptionally a higher degree of accuracy is desired, then the penalty shifting method of WIERZBICKI (see [479]) allows us to improve the results based on a low penalty level (cf. [182, Chap.6]).
- Frequently, the constraints in D are only active, for regularized solutions, if the regularization parameter $\alpha$ is extremely small

(oscillating solutions) or extremely large (oversmoothing). Exterior penalty function methods do not change the objective functional in the interior of D. Consequently, if $\alpha$ is appropriately chosen and $x_{(\alpha)} \in \mathrm{int}(D)$, then a good problem condition caused by the uniformly convex stability term $\alpha \cdot \Omega(x)$ is preserved when exterior penalty functions are added.

- After having applied quadratic exterior penalty functions to problem (6.1) the obtained new objective functional is of norm square form again.

Now let us present in a brief form the mathematical details of the numerical solution strategy concerning (6.1). The reader may consult [209, Sec.4.2] for a more refined treatment of the problem. We shall use the Cholesky decomposition $\Phi = L \cdot L^T$ (see formula (4.69)) and a representation of the convex set D by means of l continuously differentiable convex functionals $g_i(x)$ as follows:

$$D := \{ x \in R^n : \quad g_i(x) \leq 0 \ , \ i=1(1)l \} \ . \tag{6.3}$$

The Euclidean norm square character of the objective functional of (6.1) becomes evident if we write $\|L^T(x-\bar{x})\|^2$ instead of $(\Phi(x-\bar{x}), x-\bar{x})$ . Therefore, we search for optimization routines that are able to exploit the norm square form of the functional to be minimized. Moreover, define penalty functions

$$p_i(x) := \max \ (g_i(x), 0) \ , \ i=1(1)l \tag{6.4}$$

and a quadratic loss functional

$$\bar{P}(x) := \|p(x)\|^2 \quad , \quad p(x) := (p_1(x), \ldots, p_l(x))^T \ . \tag{6.5}$$

In addition to the introduction of penalty terms, the transformation of (6.1) into an unconstrained optimization problem also requires the extension of the residual norm square $\|Ax-z\|^2$ to the whole space $R^n$. This needs extending the operator A to $R^n$. In order to preserve the continuity of the objective functional, the extension of A should be continuous. In the sequel, let $\tilde{A}: R^n \rightarrow R^m$ be a continuous extension of A to the whole space $R^n$. For example, one can choose the variant

$$\tilde{A}\, x := \begin{cases} A\, x & \text{if } x \in D \\ A\, \mathbb{P}_D\, x & \text{if } x \notin D \end{cases} \tag{6.6}$$

with $\|\mathbb{P}_D x - x\| = \min\limits_{x' \in D} \|x - x'\|$. Unfortunately, the nondifferentiable character of this extension may genarate some additional numerical difficulties. Moreover, if D has a complicated structure, then the projection $\mathbb{P}_D$ is difficult to verify. In such a case, one could try to construct an extension $\tilde{A}$ of A with continuous Fréchet derivative $A'(x)$ for all $x \in R^n$. If we fail in finding a differentiable extension, we should at least exploit the fact that the natural domain $\tilde{D}$ of A is often substantially larger than D. For example, initial-boundary value problems to the heat equation, the solution of which is $Ax$, may be solvable for all nonnegative heat conductivity and heat transfer values gathered into the vector x, whereas D only considers vectors x with special properties (e.g. monotonicity behaviour of the components). If, furthermore, $\tilde{D}$ is of simple structure (preferably, it should contain lower and upper bounds on the components only), then in contrast to (6.6) the formula

$$\tilde{A}\,x := \begin{cases} A\,x & \text{if } x \in \tilde{D} \\ A P_{\tilde{D}}\,x & \text{if } x \notin \tilde{D} \end{cases} \tag{6.7}$$

avoids additional gradient discontinuities of $\|\tilde{A}x - z\|^2$ at the boundary of D whenever $D \subseteq \operatorname{int}(\tilde{D})$ .

Hence, the unconstrained auxiliary problem attains the form

$$\underset{x \in R^n}{\text{minimize}} \quad F_\beta(x, \alpha)\ , \tag{6.8}$$

where $\beta > 0$ represents the penalty parameter and

$$\begin{aligned} F_\beta(x, \alpha) &:= \|\tilde{A}x - z\|^2 + \alpha \cdot \Omega(x) + \beta \cdot \bar{P}(x) = \\ &= \|\tilde{A}x - z\|^2 + \alpha \|L^T(x - \bar{x})\|^2 + \beta \cdot \|p(x)\|^2 \ . \end{aligned} \tag{6.9}$$

the objective functional.

<u>Definition 6.1:</u>

Any global solution $x_{(\alpha,\beta)}$ to the optimization problem (6.8) is called a penalty solution associated with the regularization parameter $\alpha > 0$ and with the penalty parameter $\beta > 0$. The corresponding set of all penalty solutions $x_{(\alpha,\beta)}$ is denoted by $X_{(\alpha,\beta)}$. ○

<u>Theorem 6.2:</u>

Under the assumptions stated above we have, for $\alpha > 0$,

$$\emptyset \neq \limsup_{i \to \infty} X_{(\alpha,\beta_i)} \subseteq X_{(\alpha)} \quad \text{whenever } \beta_i \to \infty \text{ as } i \to \infty\ , \tag{6.10}$$

i.e., any sequence $\{x_{(\alpha,\beta_i)}\}_{i=1}^{\infty}$ of penalty solutions with an unlimited sequence of penalty parameters $\{\beta_i\}_{i=1}^{\infty}$ possesses a subsequence that converges to a regularized solution $x_{(\alpha)} \in X_{(\alpha)}$. ○

<u>Proof:</u>

For any regularized solution $x_{(\alpha)}$ we have

$$\Omega(x_{(\alpha,\beta_i)}) \leq \frac{1}{\alpha} \cdot F_\beta(x_{(\alpha,\beta_i)}, \alpha) \leq \frac{1}{\alpha} \cdot F_\beta(x_{(\alpha)}, \alpha) = \frac{1}{\alpha} \cdot F(x_{(\alpha)}, \alpha) \tag{6.11}$$

due to $\bar{P}(x_{(\alpha)})=0$. Hence, $\{x_{(\alpha,\beta_i)}\}_{i=1}^{\infty}$ is a bounded (relatively compact) sequence in $R^n$. Now let $\lim_{j\to\infty} x_{(\alpha,\beta_{i_j})} = x'$ be the limit vector of a convergent subsequence. The relation (6.11) implies, for all j, $\beta_{i_j}\cdot \bar{P}(x_{(\alpha,\beta_{i_j})})\leq F(x_{(\alpha)},\alpha)$ and thus $\bar{P}(x')=0$. By definition of the loss functional $\bar{P}(x)$ we so derive $x'\in D$. If we exploit formula (6.11) again, it follows the inequality $F(x',\alpha)\leq F(x_{(\alpha)},\alpha)$. Moreover, $x'\in I_{(\alpha)}$ and (6.10) holds. #

<u>Remark 6.3:</u>
Under weak assumptions it can be shown that $x_{(\alpha,\beta)}\notin D$. This helps explain the name exterior penalty function method. Furthermore, we have $\bar{P}(x_{(\alpha,\beta_1)})\geq \bar{P}(x_{(\alpha,\beta_2)})$ if $\beta_1<\beta_2$. o

In order to formalize the above discussed technique, we give an algorithmic description that is aimed at computing an approximation $x_{alg}$ to a regularized solution $x_{(\alpha)}$.

<u>Algorithm 6.4:</u>

<u>Step 0.</u> Choose a strictly increasing unlimited sequence $\{\beta_i\}_{i=1}^{\infty}$ of penalty parameters and a stopping bound $\varepsilon_1>0$. Set $i:=1$ and continue.

<u>Step 1.</u> Calculate a penalty solution $x_{(\alpha,\beta_i)}$, set $x^{(i)}:=x_{(\alpha,\beta_i)}$ and continue.

<u>Step 2.</u> If $\bar{P}(x^{(i)})\leq \varepsilon_1$, then continue; otherwise set $i:=i+1$ and return to Step 1.

<u>Step 3.</u> Set $x_{alg}:=\mathbb{P}_D\, x^{(i)}$ and stop. o

We still have to discuss the numerical computation of penalty solutions. Note that the unconstrained optimization problem (6.8) is a generally nonlinear least squares problem

$$\underset{x\in R^n}{\text{minimize}}\ \|G(x)\|^2 \tag{6.12}$$

with the transformation $G: R^n \to R^{m+n+1}$ defined by

$$G(x)=\begin{pmatrix}\tilde{A}\,x - z\\ \sqrt{\alpha}\,L^T\cdot(x-\bar{x})\\ \sqrt{\beta}\,p(x)\end{pmatrix} \quad . \tag{6.13}$$

Hence, $x_{(\alpha,\beta)}$ represents a least squares solution of the overdetermined system

$$G(x) = 0 \tag{6.14}$$

of m+n+1 equations in n variables. For the numerical analysis concerning the solution of nonlinear least squares problems, the reader is referred to SCHWETLICK [394, Chap.10] (see also DAMERT et al. [81], DE VILLERS/GLASSER [89], GILL et al. [149], [271], MC KEOWN [269], ORTEGA/RHEINBOLDT [347] and STEEN/BYRNE [406]).

In [394], globally convergent modifications to the Gauss-Newton method are especially recommended for the iterative computation of the required least squares solutions. In the author's opinion, Levenberg-Marquardt variants of the Gauss-Newton method (see [291], [309]) seem to be very efficient for solving the problem (6.12) if a Goldstein quotient globalization technique (as implemented in the computer programme DNLQ of the package [298]) controls the choice of $\gamma_j$ and $\lambda_j$ determining the iteration process

$$x^{(j+1)}:=x^{(j)}-\gamma_j(\lambda_j(G'(x^{(j)}))^T G'(x^{(j)})+(1-\lambda_j)I)^{-1}(G'(x^{(j)}))^T G(x^{(j)}) \qquad (6.15)$$

of Gauss-Newton type.

Difficulties arising from large residual effects or from gradient discontinuities seldom influence the success of the minimization routine (6.15). Whenever $A'(x)$ as a submatrix of $G'(x)$ is not explicitly available, then a derivative-free fashion of Gauss-Newton iterations using finite difference approximations should be applied. Following [298] we propose to approximate $A'(x)$ by

$$A'_\tau(x) := \frac{A(x+\tau(x_i)e^{(i)})-Ax}{\tau(x_i)} , \qquad (6.16)$$

where $e^{(i)}$ denotes the i-th unit vector and $\tau(x_i)$ the stepsize of difference approximation with respect to the i-th direction. We assume to choose the stepsizes as follows:

$$|\tau(x_i)| = \tau_{rel}\cdot|x_i| + \tau_{abs} \quad , \ i=1(1)n, \quad \tau_{rel}\geq 0, \ \tau_{abs}>0 \ . \qquad (6.17)$$

## 6.3. Some Remarks on Software

Software for the regularized solution of linear inverse problems is fairly advanced (see e.g. the software package of [441]). On the other hand, if nonlinear inverse problems are to be solved, then the needed software is generally adapted to the special class under consdieration (e.g. nonlinear integral equations, identification in partial differential equations or inverse eigenvalue problems). However, paying attention primarily to the discretized fashions of inverse problems we do not have to distinguish the different kinds of background problems throughout the numerical solution process. That means, we can establish a unifying software package for the regularized solution of nonlinear discretized inverse problems of arbitrary type. Note that the intrinsic set of objective and subjective a priori information has to be considered as a substantial component of the given discretized inverse problem. The following figure illustrates the possible structure of such a unifying software complex.

Figure 6.5:

Software for discretized inverse problems

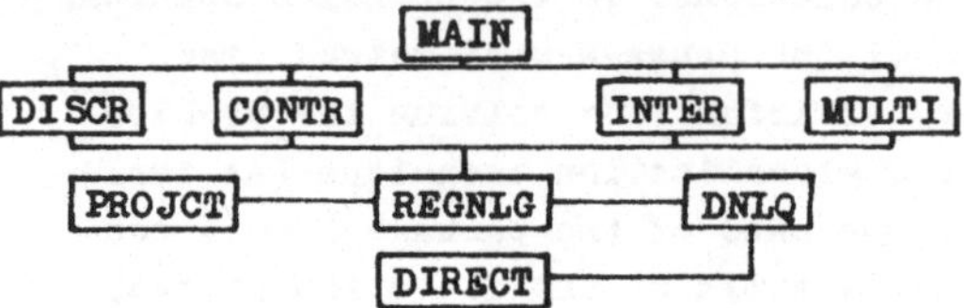


The main programme MAIN makes the input data m, n, z and bounds $\delta$ (or $\delta_1, \ldots, \delta_k$), c, $\vartheta$ and $\chi$ as well as all the values, vectors and matrices available, which express a priori information (e.g. $g_1(x)$, $\Phi$ and $\bar{x}$). Moreover, MAIN has to decide in favour of one of the four variants CONTR, DISCR, MULTI and INTER. The programme INTER is aimed at computing a finite sequence of regularized solutions. In an interactive manner, the user prescribes a regularization parameter $\alpha$ and corrects his chosen parameter value by means of the properties of the obtained regularized solutions. On the other hand, the programmes DISCR, CONTR and MULTI search for approximate regularization parameters automatically. Based on Algorithm 4.17 DISCR tries to realize the generalized discrepancy principle by computing a vector $x_{opt}$ (see problem (4.6)). The routine CONTR serves as a solver of control problems in order to determine $x_c$ (see problem (4.7)). Finally, the multi-parameter regularization according to Algorithm 4.26 is implemented in MULTI.

All four mentioned routines require an auxiliary programme for computing the solution to an optimization problem (6.1) (or to (6.2) in the case of MULTI), where a fixed regularization parameter is given. As the previous section has indicated, this auxiliary routine (called REGNLG) can be realized in a uniform manner by Algorithm 6.4. The arising unconstrained optimization problems are solved by the Gauss-Newton programme DNLQ. After achieving a sufficiently large penalty parameter $\beta$, the penalty solution will be projected into the set D by applying the programme PROJCT (see Algorithm 6.4, Step 3).

Finally, the programme DNLQ is forced to call the programme DIRECT many times. In DIRECT, the direct problem is to be solved, i.e., the range vector $\tilde{A}x$ must be calculated, for given x, where the special extension $\tilde{A}$ of A to the whole space $R^n$ is taken into account. Thus, the programme DIRECT represents the only ingredient of the presented software complex, in which the particular character of the submitted inverse problem plays an important role. The basic activities of a potential user of such a complex should be directed to an efficient implementation of the direct problem as well as to the comprehensive preparation of all required input data and a priori information. The proposed software strategy may be of interest if a computing centre has to deal with very different kinds of inverse problems. However, if the results thus obtained are not satisfactory, a refinement based on special software assigned to the considered class of inverse problems can be added.

## 6.4. An Iterative Refinement

This final section is devoted to an iterative refinement strategy for the numerical solution of discretized inverse problems. We are going to confine our discussion to identification problems in partial differential equations, where the direct problem is an initial-boundary value problem to a (linear or nonlinear) parabolic equation. It seems to be possible to apply the method also to a wider class of problems. As for numerical examples to the suggestions of this section, the reader may consult the papers [137] and [213].

Let us refer back to the strategy of computing regularized solutions $x_{(\alpha)}$ introduced in Sec. 6.2.. We a priori assume that $A: D \subseteq R^n \longrightarrow R^m$ is defined by the solution of the associated direct problem using a finite difference method on a fine grid in space and time. The application of such a fine grid ensures that the error of discretization $\|Q^m \mathcal{A} \varkappa - AP^n \varkappa\|$ is negligible compared to the error of observation whenever n is sufficiently large. However, in particular the time grid, which determines the number of time steps to be performed for the solution of one direct problem, controls the amount of computational work. If the time grid is fine, then the derivative-free Gauss-Newton iteration (6.15) subject to (6.16) and (6.17) gets very expensive. Therefore, in addition to the operator A we consider a second operator $A_0: D \subseteq R^n \longrightarrow R^m$ such that the image vector $A_0 x$ is fast computable, for any given $x \in D$. By $A_0$ we can understand an approximation of A associated with a coarse grid. For example, the number of time steps performed by a difference method may be reduced substantially in comparison with A.

Our main idea is as follows. We are going to combine the fine grid operator A and the coarse grid operator $A_o$ in order to solve the optimization problem (6.8) efficiently. Instead of applying the derivative-free Gauss-Newton method directly to problem (6.8), we solve a sequence of auxiliary problems

$$\underset{x \in R^n}{\text{minimize}} \; F_{o,\beta}^{\hat{z}}(x,\alpha) \tag{6.18}$$

with

$$F_{o,\beta}^{\hat{z}}(x,\alpha) := \|\tilde{A}_o x - \hat{z}\|^2 + \alpha\Omega(x) + \beta\cdot\bar{P}(x) \tag{6.19}$$

and various different data vectors $\hat{z}$, which express the successive adaption of the observation data z to the character of $A_o$. The solutions of (6.18) are denoted by $x_{o,(\alpha,\beta)}(\hat{z})$. Any problem (6.18) is also solved by a derivative-free Gauss-Newton method. The considerably reduced expense of the direct problem based on $A_o$ may lead to an essential reduction of the computational cost in comparison with the immediate Gauss-Newton solution of problem (6.8).

Now we will present an algorithm aimed at minimizing the total amount of work for computing a penalty solution $x_{(\alpha,\beta)}$ to problem (6.8). In the predictor part of the algorithm described below, the data vector z is adapted to the coarse grid operator $A_o$ by a sequence of difference vectors $\tilde{A}x-\tilde{A}_o x$, where x runs through different solutions $x_{o,(\alpha,\beta)}(\hat{z})$. By setting

$$z^{(0)}:=z;\; x^{(i)}:=x_{o,(\alpha,\beta)}(z^{(i-1)});\; z^{(i)}:=z-\tilde{A}x^{(i)}+\tilde{A}_o x^{(i)},\; i=1,2,\ldots \tag{6.20}$$

a predictor iteration process is prescribed. This predictor iteration must be controlled so that the sequence $\{x^{(i)}\}_{i=1}^{\infty}$ yields a good approximation $x_{pr}$ to $x_{(\alpha,\beta)}$. If it is necessary, then the vector $x_{pr}$ can further get improved by a small number of additional direct Gauss-Newton iteration steps applied to $F_\beta(x,\alpha)$ in the corrector part of the algorithm.

Algorithm 6.6:

Predictor part:

Step 0. Choose real positive stopping bounds $\varepsilon_1$, $\varepsilon_2$, $\varepsilon_3$ and a maximum number $i_{pr}$ of predictor iterates. Furthermore, choose a starting vector $x_{pr}^{(0)} \in R^n$ for the Gauss-Newton iteration in order to obtain the initial predictor iterate $x^{(1)}$ and continue.

Step 1. Set $z^{(0)}:=z$ and calculate $x^{(1)}:=x_{o,(\alpha,\beta)}(z^{(0)})$ by solving (6.18) based on a derivative-free Gauss-Newton iteration. Set $s_1:=F_\beta(x^{(1)},\alpha)$, $x_{pr}:=x^{(1)}$, $s:=s_1$, $j:=0$, $i:=1$ and continue.

**Step 2.** If $i=i_{pr}$, then go to Step 5; otherwise set $z^{(i)}:=z-\tilde{A}x^{(i)}+\tilde{A}_0x^{(i)}$. If we have $\|z^{(i)}-z^{(i-1)}\|\leq\varepsilon_1$, then go to Step 5, otherwise calculate $x^{(i+1)}:=x_{0,(\alpha,\beta)}(z^{(i)})$ by solving (6.18) based on a derivative-free Gauss-Newton iteration and continue.

**Step 3.** If $\|x^{(i+1)}-x_{pr}\|\leq\varepsilon_2$, then go to Step 5; otherwise set $s_{i+1}:=F_\beta(x^{(i+1)},\alpha)$. If we have the inequality $s_{i+1}>s-\varepsilon_3$, then go to Step 4; otherwise set $x_{pr}:=x^{(i+1)}$, $s:=s_{i+1}$, $i:=i+1$, $j:=0$ and return to Step 2.

**Step 4.** If $j=1$, then go to Step 5; otherwise set $i:=i+1$, $j:=1$ and return to Step 2.

Corrector part:

**Step 5.** Choose a maximum number $i_{corr}$ of admissible corrector iterates. Use the vector $x^{(0)}_{corr}:=x_{pr}$ as a starting approximation for the Gauss-Newton iteration (6.15) applied to problem (6.8) and continue.

**Step 6.** Improve the initial guess $x^{(0)}_{corr}$ by performing a few corrector iteration steps, which are directly aimed at minimizing $F_\beta(x,\alpha)$. o

For the further explanation of the predictor part, we are going to add some remarks. In Step 0, the stopping bounds which control the predictor iteration (6.20) are chosen. The corresponding stopping criteria for the inner Gauss-Newton iterations according to a fixed integer i are not mentioned explicitly in Algorithm 6.6. However, Theorem 6.10 will refer to approximate solutions of problem (6.18) and their influence to the quality of the vector $x_{pr}$. There are, on the other hand, four reasons for terminating the predictor iteration. Firstly, the iteration number is bounded by $i_{pr}$. Secondly, no improvement of the solution can be expected if the data are almost not changed. Therefore, the process terminates if an inequality $\|z^{(i)}-z^{(i-1)}\|\leq\varepsilon_1$ is valid. The predictor part also comes to an end if a new iterate is close to the best of all approximations, which are already known, i.e., $\|x^{(i+1)}-x_{pr}\|\leq\varepsilon_2$. This criterion becomes active if the sequence $\{x^{(i)}\}_{i=1}^\infty$ is converging to a limit vector $x^{(\infty)}$ as assumed below in Theorem 6.7. In such a case, we generally have $F_\beta(x_{(\alpha,\beta)},\alpha)<F_\beta(x^{(\infty)},\alpha)$ and the predictor iteration is aimed at finding the minimizer $x_{pr}$ of $F_\beta(x^{(i)},\alpha)$ over all i. The predictor iteration also stops if two additionally computed predictor iterates do not reduce the obtained minumum value s of

$F_\beta$ at least by an amount of $\varepsilon_3$. In Step 1, the initial predictor iterate $x^{(1)}$ is verified with unadapted data. However, the following iterates $x^{(i)}$, $i=2,3,\dots$ always require adapted data. If the (i+1)-th iterate fails to reduce $F_\beta$ sufficiently, then by command of Step 3 the algorithm checks in Step 4 whether j=0 or j=1 holds. In the case j=0, the i-th predictor iterate has been computed by a normal predictor iteration step, i.e., $F_\beta$ has been reduced by a minimum amount of $\varepsilon_3$. Then, by setting j:=1 an auxiliary step will be performed. If otherwise j=1, then the algorithm terminates, since the auxiliary step has already failed.

The following theorem motivates the predictor iteration of Algorithm 6.6.

<u>Theorem 6.7:</u>
Assume that $\tilde{A}: R^n \to R^m$ and $\tilde{A}_o: R^n \to R^m$ are extensions of A and $A_o$ to $R^n$ with continuous Fréchet derivatives $\tilde{A}'(x)$ and $\tilde{A}_o'(x)$ on the whole space $R^n$. Furthermore, let $\{x^{(i)}\}_{i=1}^{\infty}$ be an infinite sequence of iterates satisfying (6.20). If this sequence is converging to a limit vector $x^{(\infty)}=\lim_{i\to\infty} x^{(i)}$, then we have

$$\|\nabla_x F_\beta(x^{(\infty)},\alpha)\| \leq 2\cdot\|\tilde{A}'(x^{(\infty)})-\tilde{A}_o'(x^{(\infty)})\|\cdot\|\tilde{A}x^{(\infty)}-z\| \tag{6.21}$$

and

$$\begin{aligned}\|\nabla_x F_\beta(x^{(1)},\alpha)\| \leq 2\cdot\{&\|\tilde{A}'(x^{(1)})-\tilde{A}_o'(x^{(1)})\|\cdot\|\tilde{A}x^{(1)}-z\|+\\ &+\|\tilde{A}_o'(x^{(1)})\|\cdot\|\tilde{A}x^{(1)}-\tilde{A}_o x^{(1)}\|\} \;.\;\circ\end{aligned} \tag{6.22}$$

<u>Proof:</u> [137, Appendix].

Due to the data adaption process, the influence of differences $\tilde{A}x-\tilde{A}_o x$ disappears in formula (6.21). Provided that the residual norms $\|\tilde{A}x-z\|$ as well as the Fréchet derivative differences $\|\tilde{A}'(x)-\tilde{A}_o'(x)\|$ are of the same order of magnitude for both vectors $x^{(\infty)}$ and $x^{(1)}$, the gradient norm $\|\nabla_x F_\beta(x^{(\infty)},\alpha)\|$ is expected to be smaller than $\|\nabla_x F_\beta(x^{(1)},\alpha)\|$. Under the above stated assumptions there is a good chance that, in comparison with the vector $x^{(1)}:=x_{o,(\alpha,\beta)}(z)$, appropriate predictor iterates represent considerably improved approximations to $x_{(\alpha,\beta)}$. However, existence and quality of $x^{(\infty)}$ substantially depend on derivative deviations $\|\tilde{A}'(x)-\tilde{A}_o'(x)\|$. Numerical experience according to inverse problems in parabolic equations indicates that the error influence of difference terms $\tilde{A}x-\tilde{A}_o x$ to the solution is often large with respect to the influence of derivative differences.

**Assumption 6.8:**

In the sequel, let the assumptions of Theorem 6.7 hold. In addition to these assumptions suppose that the problems (6.18) have a unique solution $x_{0,(\alpha,\beta)}(\hat{z})$, for all vectors $\hat{z}$ of a closed sphere $S_z=\bar{S}(z,c_1)$. Furthermore, let hold a Lipschitz condition:

$$\|x_{0,(\alpha,\beta)}(\hat{z}^{(1)})-x_{0,(\alpha,\beta)}(\hat{z}^{(2)})\| \leq c_0\cdot\|\hat{z}^{(1)}-\hat{z}^{(2)}\| , \quad c_0>0, \tag{6.23}$$

for all $\hat{z}^{(1)},\hat{z}^{(2)}\in S_z$. Moreover, we postulate the inequalities

$$\begin{aligned} &\|\tilde{A}x-\tilde{A}_0x\| \leq c_1 , \\ &\|\tilde{A}'(x)-\tilde{A}_0'(x)\|\leq c_2 , \end{aligned} \tag{6.24}$$

for all vectors $x$ of a closed sphere $S_x=\bar{S}(x_{0,(\alpha,\beta)}(z),c_3)$, where

$$c_0\cdot c_1 \leq c_3 \tag{6.25}$$

is supposed. ○

**Theorem 6.9:**

Under the assumptions stated above the operator $B$: $S_x\subseteq R^n\longrightarrow R^n$, which is uniquely defined by the formula

$$Bx := x_{0,(\alpha,\beta)}(z-\tilde{A}x+\tilde{A}_0x) , \quad x\in S_x , \tag{6.26}$$

maps the sphere $S_x$ in itself. Whenever

$$q:= c_0\cdot c_2<1 , \tag{6.27}$$

then (6.25) describes a contraction mapping and there exists a uniquely determined fixed point $x^{(\infty)}\in S_x$ satisfying the equation $Bx^{(\infty)} = x^{(\infty)}$ and the inequality (6.21). Furthermore, for a sequence $\{x^{(i)}\}_{i=1}^{\infty}$ according to (6.20), the inequality (6.27) implies

$$\begin{aligned} &\|x^{(i)}-x^{(\infty)}\| \leq q\cdot\|x^{(i-1)}-x^{(\infty)}\|, \quad i=2,3,\ldots , \\ &\|x^{(i)}-x^{(\infty)}\| \leq \frac{q^{i-1}}{1-q}\cdot c_0\cdot c_1 \quad . \ ○ \end{aligned} \tag{6.28}$$

**Proof:**

The first inequality of (6.24) ensures that, for $x\in S_x$, $z-\tilde{A}x+\tilde{A}_0x\in S_z$. Furthermore, for all $x,\tilde{x}\in S_x$ we have $\|Bx-B\tilde{x}\|= \|x_{0,(\alpha,\beta)}(z-\tilde{A}x+\tilde{A}_0x)-x_{0,(\alpha,\beta)}(z-\tilde{A}\tilde{x}+\tilde{A}_0\tilde{x})\|\leq c_0\cdot\|(\tilde{A}-\tilde{A}_0)x-(\tilde{A}-\tilde{A}_0)\tilde{x}\|\leq c_0\cdot c_2\cdot\|x-\tilde{x}\|$. If the inequality (6.27) is valid, this implies $\|Bx-B\tilde{x}\|\leq q\cdot\|x-\tilde{x}\|$, for all $x,\tilde{x}\in S_x$ and $q<1$. Hence, $B$ is a contracting operator on $S_x$. On the other hand, in view of formula (6.25) we have

$\|x_{0,(\alpha,\beta)}(z)-x_{0,(\alpha,\beta)}(z-\tilde{A}x+\tilde{A}_0x)\|\leq c_0\|\tilde{A}x-\tilde{A}_0x\|\leq c_0\cdot c_1 \leq c_3$

whenever $x\in S_x$. Consequently, $Bx\in S_x$ if $x\in S_x$. The operator $B$ maps the sphere $S_x$ in itself. Thus, the assumptions of Banach's fixed point theorem are satisfied (see e.g. COLLATZ [70, Chap.12]). Following the assertions of that theorem, a fixed point $x^{(\infty)}$ with $Bx^{(\infty)}=x^{(\infty)}$ exists in $S_x$ and is uniquely determined in this set. Moreover, a sequence $\{x^{(i)}\}_{i=1}^{\infty}$ according to (6.20) fulfils the inequalities (6.28). ○

There are good chances that the predictor iteration sequence $\{x^{(i)}\}_{i=1}^{\infty}$ of Algorithm 6.6 converges to a vector $x^{(\infty)}$ whenever the Fréchet derivative majorant $c_2$ of the operator difference $\tilde{A}-\tilde{A}_0$ is small enough. In such a case, the predictor iteration is linearly convergent as formula (6.28) emphasizes. Note that the obtained qualitative results remain true if we only know approximate solutions to the problem (6.18) (cf. also ALFELD [6]).

Theorem 6.10:

Let $\{\hat{x}^{(i)}\}_{i=1}^{\infty} \subseteq S_x$ be a sequence of approximations to the sequence $\{x^{(i)}\}_{i=1}^{\infty}$ formed by (6.20) such that

$$\|\hat{x}^{(i)}-x_{0,(\alpha,\beta)}(\hat{z}^{(i-1)})\| \leq \varepsilon , \tag{6.29}$$

$$\hat{z}^{(0)}:= z , \quad \hat{z}^{(i)}:= z-\tilde{A}\hat{x}^{(i)}+\tilde{A}_0\hat{x}^{(i)}, \quad i=1,2,\ldots .$$

Provided that the assumptions of Theorem 6.9 hold, we have

$$\|\hat{x}^{(i)}- x^{(i)}\| \leq \frac{\varepsilon}{1-q} \quad . \quad ○ \tag{6.30}$$

Proof:

The proposition (6.30) of this theorem is a consequence of Banach's fixed point theorem applied to perturbed operators (see [70,p.174]). #

The iterative regularized solution of nonlinear discretized identification problems in parabolic equations is often very expensive, especially if Fréchet derivatives $A'(x)$ of A are to be approximated by differences of vectors Ax. Whenever the proposed Algorithm 6.6 proceeds satisfactory, an approximate operator $A_0$ to A associated with a coarse grid in time discretization and fast computable vectors $\tilde{A}_0x$ can help to minimize the total amount of work for the solution of the inverse problem. Computational experience concerning inverse heat equation problems points out that the predictor part of the algorithm is very suitable in order to reduce the expense of the starting phase in the minimization of $F_\beta(x,\alpha)$. The two-level method seems to be more efficient than continuation strategies (see DE VILLERS/GLASSER [89]) exploiting one level only.

There are still many open questions with respect to the selection of $A_o$ as well as with respect to the choice of appropriate stopping criteria for the inner iteration aimed at minimizing (6.18). The paper [213] studies, in detail, the influence of the additional discretization error arising from the use of $A_o$ instead of A to the solution of the parabolic inverse problem. A further justification for the combined application of two operator levels A and $A_o$ is given based on values of the modulus functional (3.52) (see Sec. 3.3.). The generalization of the introduced technique to the multi-level case will open an additional way to reduce the computing costs whenever an efficient numerical control regime can be made available. However, a lot of theoretical and experimental work is still to be done in this field.

**NOTATIONS**

| | |
|---|---|
| $\circ$ | the end of a theorem, lemma, corollary, definition, remark or algorithm |
| # | the end of a proof |
| $R$, $R^n$, $R^m$ | finite dimensional real Euclidean vector spaces of dimensions 1, n and m, respectively |
| $R^{m\times n}$ | space of real m×n-matrices |
| $B_o$, $B_1$, $B_2$ | real Banach spaces |
| $H$, $H_1$, $H_2$, $H_3$ | real Hilbert spaces |
| $[t_1,t_2]$ | closed interval $t_1\le t\le t_2$ in R |
| $[t_1,\infty)$ | half axis $t_1\le t<\infty$ in R |
| $C[t_1,t_2]$ | Banach space of continuous real functions over $[t_1,t_2]$ |
| $L_2[t_1,t_2]$ | Hilbert space of quadratically integrable real functions over $[t_1,t_2]$ |
| $C^{(1)}[t_1,t_2]$ | Banach space of continuously differentiable real functions over $[t_1,t_2]$ |
| $L_p[t_1,t_2]$ | Banach space of p-th power integrable real functions over $[t_1,t_2]$ |
| $W_p^{(k)}[t_1,t_2]$ | Sobolev space of functions over $[t_1,t_2]$ possessing p-th power integrable generalized derivatives up to the k-th order |
| $\mathring{W}_p^{(k)}[t_1,t_2]$ | subspace of $W_p^{(k)}[t_1,t_2]$ with homogeneous boundary data |
| $H_o^{(\mu)}[t_1,t_2]$ | spaces of a continuous Hilbert scale (see Chap.2) |
| $H^{(\mu)}$ | space of special tempered distributions (see Chap.2) |
| $C^{(\infty)}(P)$ | Banach space of infinitely continuously differentiable functions over a domain P |
| $\Vert.\Vert$, $(.,.)$ | Euclidean norm and inner product in finite dimensional spaces |
| $\Vert.\Vert_*$ | additional norm in finite dimensional spaces |
| $\Vert.\Vert_i$ | norm in a Banach space $B_i$ |
| $(.,.)_i$ | inner product in a Hilbert space $H_i$ |
| $s,t,\alpha,\beta,\gamma,\delta\ldots$ | real numbers |
| $x$, $b$ | real vectors $x=(x_1,x_2,\ldots,x_n)^T$, $b=(b_1,b_2,\ldots,b_m)^T$ |
| $x_i$, $b_j$ | real vector components |
| $x^T$ | transposed vector to x |
| $x(t)$, $b(s)$ | real functions of a real variable |
| $x'(t)$, $x''(t)$ | first and second derivative of $x(t)$ |
| $x^{[\nu]}(t)$ | $\nu$-th derivative of $x(t)$ |
| $\varkappa$, $\mathcal{b}$ | Banach space elements |
| $\varkappa=\langle x(t),\ t_1\le t\le t_2\rangle$ | Banach space element associated with a real function over an interval $[t_1,t_2]$ |
| $f(t)\propto g(t)$ | f is a function proportional to g |
| $\{x^{(i)}\}_{i=1}^{k}$ | finite sequence of vectors $x^{(1)},\ldots,x^{(k)}$ |

| | |
|---|---|
| $\{x^{(i)}\}_{i=1}^{\infty}$ | infinite sequence of vectors |
| $\{\varkappa^{(i)}\}_{i=1}^{\infty}$ | infinite sequence of Banach space elements |
| $\mathcal{D}$ | closed convex subset of a Banach space $B_1$ |
| $D$ | closed convex subset of $R^n$ |
| $\mathcal{A}: \mathcal{D} \subseteq B_1 \rightarrow B_2$ | continuous operator from $B_1$ into $B_2$ |
| $A: D \subseteq R^n \rightarrow R^m$ | continuous operator from $R^n$ into $R^m$ |
| $\\|\mathcal{A}\\|$ | norm of a linear operator $\mathcal{A}$ |
| $\mathcal{E}$ | embedding operator |
| $0$ | zero, zero vector, zero element |
| $I$ | unity matrix, unity operator |
| $P^n: B_1 \rightarrow R^n$ | projector of a priori discretization |
| $Q^m: B_2 \rightarrow R^m$ | projector of experimental design |
| $\rho^n: D \subseteq R^n \rightarrow B_1$ | back projector |
| $A=(a_{ji})_{i=1(1)n}^{j=1(1)m}$ | m×n-matrix with entries $a_{ji}$ |
| $A^T$ | transposed matrix to A |
| $\\|A\\|$, $\\|A\\|_F$ | spectral and Frobenius norms of A |
| $\mathcal{A}^{-1}$ | inverse operator to $\mathcal{A}$ |
| $\mathcal{A}^{(-1)}$ | multi-valued inverse mapping to $\mathcal{A}$ |
| $\mathcal{A}_L^{-1}$ | left inverse to a linear operator $\mathcal{A}$ |
| $A^+$ | Moore-Penrose inverse to a matrix A |
| $\mathcal{A}\mathcal{D}$, $AD$ | range subsets $\{\ell = \mathcal{A}\varkappa,\ \varkappa \in \mathcal{D}\}$ and $\{b=Ax,\ x \in D\}$ |
| $\mathcal{A}\varkappa$, $Ax$ | image element and image vector of $\mathcal{A}$ and A applied to $\varkappa$ and x |
| $\mathbb{N}(\mathcal{A})$, $\mathbb{N}(A)$ | null-space of a linear operator |
| $\mathbb{R}(\mathcal{A})$, $\mathbb{R}(A)$ | range of a linear operator |
| $\mathcal{X}(\ell)$ | set $\{\varkappa \in \mathcal{D}: \\|\mathcal{A}\varkappa - \ell\\|_2 = \inf_{\varkappa' \in \mathcal{D}} \\|\mathcal{A}\varkappa' - \ell\\|_2\}$ |
| $X(b)$ | set $\{x \in D: \\|Ax-b\\| = \inf_{x' \in D} \\|Ax'-b\\|\}$ |
| $\{\ldots\}$ | set |
| $\underset{\varkappa \in \mathcal{D}}{\text{minimize}}\ \ell(\varkappa)$<br>$\underset{x \in D}{\text{minimize}}\ f(x)$<br>$\underset{x \in D,\ \\|Ax-z\\| \le \delta}{\text{minimize}}\ f(x)$ | minimization problems aimed at minimizing an objective functional $\ell$ or f over the sets of feasible solutions $\mathcal{D}$, D and $\{x \in D: \\|Ax-z\\| \le \delta\}$ |
| $\underset{x \in R^n}{\text{minimize}}\ f(x)$ | unconstrained minimization problem |
| $\text{dist}(\varkappa, \mathcal{X})$ | distance of an element $\varkappa$ from a set $\mathcal{X}$ |
| $\text{qdist}(\mathcal{X}, \mathcal{Y})$ | quasidistance between two sets $\mathcal{X}$ and $\mathcal{Y}$ |
| $R_+^k$ | $\{x \in R^k: x_i \ge 0,\ i=1(1)k\}$ |

| | |
|---|---|
| trace(A) | trace of a matrix A |
| dim(H) | dimension of a space H |
| rank(A) | rank of a matrix A |
| int(D) | interior of a set D |
| span(...) | linear hull |
| supp(x) | support of a function x(t) |
| const. | constant value or vector |
| iff | if and only if |
| $\bar{T}$ | closure of a set T |
| $\bar{H}(\mathfrak{D})$ | closure of the enveloping linear manifold of $\mathfrak{D}$ |
| $2^{\mathfrak{D}}$ | power set of $\mathfrak{D}$ |
| $\times$ | Cartesian product |
| $\oplus$ | orthogonal sum |
| O(.), o(.) | Landau symbols |
| $\mathfrak{B}$ | unbounded linear operator with maximum domain $\mathcal{T}(\mathfrak{B})$ |
| $\mathfrak{A}^{\vartheta}$ | $\vartheta$-th power of a linear operator $\mathfrak{A}$ in a Hilbert space |
| $\Pi(\varkappa)$, $\Omega(x)$ | real functionals |
| $\ell(\varkappa)$ | linear functional |
| exp(t) | $e^t$ |
| $A'(x)$ | Fréchet derivative of A at x |
| $\hat{\mathfrak{x}}$, $\hat{x}(z)$ | estimate and estimator |
| $\xi, \eta, \zeta, \ldots$ | random vectors with realizations x, y, z, ... |
| $\xi \sim \mathcal{N}(\bar{x}, B)$ | $\xi$ is Gaussian distributed with mean vector $\bar{x}$ and covariance matrix B |
| $cond_i(.)$ | condition numbers |
| $W_c^{\Pi}$, $W_c^{\Omega}$ | level sets (see Chap.3) |
| $\emptyset$ | the empty set |
| $\rightarrow$, $\xrightarrow[1]{}$ | strong convergence in $R^n$ and $B_1$ |
| $\xrightarrow[1]{}$ | weak convergence in $B_1$ |
| $\limsup_{i\rightarrow\infty}$ | superior limit |
| E (...) | expectation value or vector |
| P(...) | probability |
| $\text{p-lim}_{k\rightarrow\infty}$ | stochastic limit |
| $p_\eta(.)$ | density functional of a random vector $\eta$ |
| $\mathbb{P}_D$ | Euclidean norm projection into a set D |
| $\min(t_1,t_2)$ | minimum of two real numbers |
| $\max(t_1,t_2)$ | maximum of two real numbers |
| X, Z, S, T | subsets of a finite dimensional space |
| $\mathcal{X}, \mathcal{Z}, \mathcal{K}, \mathcal{U}$ | subsets of Banach spaces |
| $\nabla$, $\nabla_x$ | gradient and gradient with respect to x |

## BIBLIOGRAPHY

**Part I: Proceedings of Congresses and Seminars, Collections of Papers**


[C1] LAVRENTEV, M.M. and ALEXEEV, A.S. (eds.): Mathematical problems in geophysics. Akad. Nauk SSSR, Sibir. otd., Comput. Centre, Vyp 4, Novosibirsk 1973. In Russian.

[C2] Ill-posed problems I and II. Izd.-vo Nauka, Moscow 1974. In Russian.

[C3] Solution methods for conditional well-posed problems. Trudy Inst. Mat. i Mekh. Uralski Nauchni Zentr, Akad. Nauk SSSR, Vyp 17, Sverdlovsk 1975. In Russian.

[C4] IVANOV, V.K. (ed.): Regularization methods for instable problems. Akad. Nauk SSSR, Uralski Nauchni Zentr, Sverdlovsk 1976. In Russian.

[C5] BJÖRCK, A. (ed.): Symposium on mathematical and numerical analysis of inverse and ill-posed problems. Linköping University. Sweden 1977.

[C6] LAVRENTEV, M.M. (ed.): Nonclassical methods in geophysics. Akad. Nauk SSSR, Sibir. otd., Comput. Centre, Novosibirsk 1977. In Russian.

[C7] SAMARSKI, A.A. (ed.): Problems of mathematical physics and numerical mathematics. Izd.-vo Nauka, Moscow 1977. In Russian.

[C8] LAVRENTEV, M.M. (ed.): Inverse problems for differential equations of mathematical physics. Akad. Nauk SSSR, Sibir. otd., Comput. Centre, Novosibirsk 1978. In Russian.

[C9] RAY, W.H. and LAINIOTIS, D.G. (eds.): Distributed parameter systems - identification, estimation and control. M. Dekker, New York 1978.

[C10] SABATIER, P.C. (ed.): Applied inverse problems. Springer, Berlin-Heidelberg-New York 1978.

[C11] Distributed parameter systems - modelling and identification. Proc. IFIP Working Conference Roma, Springer, Berlin-Heidelberg-New York.1978.

[C12] Die moderne Potentialtheorie als Grundlage des inversen Problems in der Geophysik. Geodät. u. geophys. Veröff., Reihe III, no.45, Freiberg 1978.

[C13] ANGER, G. (ed.): Inverse and improperly posed problems in differential equations. Proc. Conf. Halle, Akademie-Verlag, Berlin 1979.

[C14] HERMAN, G.T. (ed.): Image reconstruction from projections. Springer, Berlin-Heidelberg-New York 1979.

[C15] LAVRENTEV, M.M. (ed.): Conditional well-posed problems in geophysics. Akad. Nauk SSSR, Sibir. otd., Comput. Centre, Novosibirsk 1979. In Russian.

[C16] LAVRENTEV, M.M. (ed.): Ill-posed mathematical problems in geophysics. Akad. Nauk SSSR, Sibir. otd., Comput. Centre, Novosibirsk 1979. In Russian.

[C17] ANDERSSEN, R.S. and DE HOOG, F.R. and LUKAS, M.A. (eds.): The application and numerical solution of integral equations. Sijthoff and Noordhoff, Alphen aan den Rijn 1980.

[C18] LAVRENTEV, M.M. (ed.): Uniqueness, stability and methods for inverse and ill-posed problems. Akad. Nauk SSSR, Sibir. otd., Comput. Centre, Novosibirsk 1980. In Russian.

[C19] HERMAN, G.T. and NATTERER, F. (eds.): Mathematical aspects of computerized tomography. Springer, Berlin-Heidelberg-New York 1981.

[C20] Direct and inverse problems in scattering theory. Inst. Math. Akad. Nauk SSSR, Uralski otd., Kiev 1981. In Russian.

[C21] Solution methods for ill-posed problems. In: Numerical methods and programming 35. Izd.-vo Mosk. Univ., Moscow 1981, pp. 3-81. In Russian.

[C22] BAKER, T.H. and MILLER, G.F. (eds.): Treatment of integral equations by numerical methods. Proc. Durham Conference, Academic Press, London-New York 1982.

[C23] JÄCKEL, H. (ed.): Mathematische Grundlagen zu Warmbehandlungstechnologien von Industriestählen. Wissenschaftliche Schriftenreihe der Techn. Hochschule Karl-Marx-Stadt 10/1982.

[C24] MOROZOV, V.A. (ed.): Methods and algorithms in numerical analysis. Izd.-vo Mosk. Univ., Moscow 1982. In Russian.

[C25] BAKHVALOV, N.S. et al. (eds.): Numerical analysis - methods, algorithms and programmes. Izd.-vo Mosk. Univ., Moscow 1983. In Russian.

[C26] DEUFELHARD, P. and HAIRER, F. (eds.): Numerical treatment of inverse problems in differential and integral equations. Birkhäuser, Boston 1983.

[C27] GORENFLO, R. and HOFFMANN, K.-H. (eds.): Applied nonlinear functional analysis. Variational methods and ill-posed problems. P. Lang, Frankfurt am Main- Berlin 1983.

[C28] HÄMMERLIN, G. and HOFFMANN, K.-H. (eds.): Improperly posed problems and their numerical treatment. Proc. Conf. Oberwolfach, Birkhäuser Basel-Boston-Stuttgart 1983.

[C29] MOROZOV, V.A. and SAMARIN, M.K. (eds.): Numerical methods and programming 39. Izd.-vo Mosk. Univ., Moscow 1983, pp. 4-81. In Russian.

[C30] FRIEDRICH, V. and SCHNEIDER, M. and SILBERMANN, B. (eds.): Probleme und Methoden der Mathematischen Physik. Proc. Conf. 8. TMP Karl-Marx-Stadt, Teubner, Leipzig 1984.

## Part II: Monographs and Original Papers

[1] ADAMS, R.A. (1975): Sobolev-spaces. Academic Press, New York.

[2] AKHIESER, N.I. (1967): Vorlesungen über Approximationstheorie. 2. Auflage, Akademie-Verlag, Berlin. Aus dem Russischen.

[3] AKHIESER, N.I. and GLASMANN, I.M. (1960): Theorie der linearen Operatoren im Hilbert-Raum. 3.Auflage, Akademie-Verlag, Berlin. Aus dem Russischen.

[4] ALBER, J.I. and RIAZANCEVA, I.P. (1975): Regularization of nonlinear equations with monotonous operators. Shurnal Vych. Math. i Mat. Fiz. 15, 283-289. In Russian.

[5] ALEXANDROV, P.S. (1971): Einführung in die Mengenlehre und die Theorie der reellen Funktionen. Dt. Verl. d. Wiss., Berlin 1971. Aus dem Russischen.

[6] ALFELD, P. (1982): Fixed point iteration with inexact function values. Math. Comp. 38, 87-98.

[7] ALT, H.W., HOFFMANN, K.-H. and SPREKELS, J. (1983): A numerical procedure to solve certain identification problems. Preprint no. 19, Universität Augsburg, BRD.

[8] AMOUROUX, M., BABARY, J.P. and MALANDRAKIS, C. (1976): Optimal location of sensors for linear stochastic distributed parameter systems. In: [C11], 92-113.

[9] ANDERSSEN, R.S. (1980): On the use of linear functionals for Abel-type integral equations in applications. In: [C17] , 195-221.

[10] ANDERSSEN, R.S. and BLOOMFIELD, P. (1974): Numerical differentiation procedures for non-exact data. Numerische Mathematik 22, 157-182.

[11] ANGER, G. (1980): Lectures on potential theory and inverse problems. In: [C12], 15-81.

[12] ANGER, G. and KLEINE, E. (1984): Some special inverse problems for the Laplace equation and the Helmholtz equation. Preprint no. 91, Martin-Luther-Universität Halle, Sektion Mathematik.

[13] ANTOKHIN, Y.T. (1967): Improperly posed problems in the Hilbert space and stable methods for their solution. Diff. Uravnenia 3, 1135-1156. In Russian.

[14] ARCANGELI, R. (1966): Pseudo-solution de l'équation Ax=y. C.R. Acad. Sci. Paris, Series A, 263, 282-285.

[15] ARONSZAJN, N. (1950). Theory of reproducing kernels. Transact. Amer. Math. Soc. 68, 337-404.

[16] ARSENIN, V.Y. and SYABREV, N.B. (1977): On optimal approximate solutions to singular and ill-conditioned systems of linear algebraic and operator equations. Preprint no. 127. IPM ANSSSR, Moscow. In Russian.

[17] ATHANS, M. (1972): On the determination of optimal costly measurement strategies for linear stochastic systems. Automatica 8, 397-412.

[18] BAEV, A.V. (1979): On the construction of normal solutions for nonlinear ill-posed problems by regularization. Shurnal Vych. Math. i Mat. Fiz. 19, 594-600. In Russian.

[19] BAGLAI, P.D. (1975): On criteria for choosing regularization parameter based on the sensitivity function. Shurnal Vych. Mat. i Mat. Fiz. 15, 305-320.

[20] BAKER, C.T.H. (1974): Methods for Volterra equations of first kind. In: Numerical solution of integral equations (eds. L.M. Delves and J. Walsh). Clarendon Press, Oxford, 162-174.

[21] BAKUSHINSKI, A.B. (1979): On the principle of iterative regularization. Shurnal Vych. Mat. i Mat. Fiz. 19, 1040-1043.

[22] BALAKRISHNAN, A.V. (1975): Identification-inverse problems for partial differential equations. In: Lecture Notes in Comp. Sci. 27, Springer Berlin-Heidelberg-New York, 1-12.

[23] BALAKRISHNAN, A.V. (1978): Identification of distributed parameter systems - Non computational aspects. In [C11] , 1-10.

[24] BANKS, H.T. and KUNISCH, K. (1982): An approximation theory for nonlinear partial differential equations with applications to identification and control. SIAM J. Control and Opt. 20, 815-849.

[25] BARCILON, V. and TURCHETTI, G. (1979): On an inverse eigenvalue problem with truncated spectral data. In: [C13], 25-34.

[26] BARD, Y. (1974): Nonlinear parameter estimation. Academic Press New York.

[27] BATES, D.M. and WAHBA, G. (1982): Computational methods for generalized cross-validation with large data sets. In: [C22], 283-296.

[28] BENSOUSSAN, A. (1971): Filtrage optimal des systémes linéaires. Dunod Paris.

[29] BERESIN, I.S. and SHIDKOV, N.P. (1970): Numerische Methoden. VEB Dt. Verlag d. Wissenschaften Berlin. Aus dem Russischen.

[30] BERSENEV, S.M. (1984): On computational schemes to the regularization method. Shurn. Vych. Mat. i Mat. Fiz. 24, 1402-1405.

[31] BERTSEKAS, D.P. (1980): Penalty and multiplier methods. In: Nonlinear optimization (eds. L.C.W. Dixon et al.). Birkhäuser Boston, 253-278.

[32] BJÖRCK, A. (1981): Least squares methods in physics and engineering. CERN 81-16, European Org. for Nuclear Research, Geneva, Switzerland.

[33] BJÖRCK, A. and ELDÉN, L. (1979): Methods in numerical algebra for ill-posed problems. Preprint LiTH-MAT-R-33-1979, Linköping Univ., Dept. of Mathematics, Sweden.

[34] BJÖRCK, A. and ELFVING, T. (1979): Accelerated projection methods for computing pseudoinverse solutions of systems of linear equations. BIT 19, 145-163.

[35] BOCK, H.G. (1981): Numerical treatment of inverse problems in chemical reaction kinetics. In: Modelling of chemical reactions systems (eds. K.H. Ebert et al.). Springer Berlin, 102-125.

[36] BOCK, H.G. (1983): Recent advances in parameter identification techniques for o.d.e. In [C26], 95-121.

[37] BOLT, B.A. (1980): What can inverse theory do for applied mathematics and sciences. Austral. Math. Soc. Gaz. 7, 69-78.

[38] BRÄUNINGER, J. (1981): A variable metric algorithm for unconstr. minimization without evaluation of derivatives. Numer. Math. 36, 359-373.

[39] BROWN, A.L. and PAGE, A. (1970): Elements of functional analysis. Van Nostrand Reinhold Comp. London.

[40] BUNKE, H. (1980): Parameter estimation in nonlinear regression models. Handbook of Statistics I (ed. P.R. Krishnaiah), North-Holland Publ. Comp. 593-615.

[41] BUNKE, H. (1981): A note on parameter estimation in inadequate nonlinear regression models. Math. Operationsforsch. u. Statist. Series Statistics 12, 7-11.

[42] BUNKE, H. (1981): Parameterschätzung in nichtlinearen Regressionsmodellen. Berichte der Math. Gesell. DDR- Mathematikerkongreß, Vortragsauszüge no 1, 11-30.

[43] BUNKE, O. (1973): Model choice and parameter estimation in regression analysis. Math. Operationsforsch. u. Statistik 4, 407-423.

[44] BUNKE, O. (1977): Mixed models, empirical Bayes and Stein estimators Math. Operationsforsch. u. Statistik, Series Statistics 8, 55-68.

[45] BUNKE, H. and O. (1974): Das empirische Entscheidungsprinzip und die Wahl von Regressionsmodellen. Biometrische Zeitschrift 16, 167-184.

[46] BUNKE, H. and GLADITZ, J. (1974): Empirical linear Bayes decision rules for a sequence of linear models with different regressor matrices. Math. Operationsforsch. u. Statistik 5, 235-244.

[47] BUNKE, H. and SCHMIDT, W.H. (1980): Asymptotic results on nonlinear approximation of regression functions and weighted least-squares. Math. Operationsforsch. u. Statistik, Series Statistics 11, 3-22.

[48] BUTLER, J.P., REEDS, J.A. and DAWSON, S.V. (1981): Estimating solutions of first-kind equations with nonnegative constraints and optimal smoothing. SIAM J. Numer. Anal. 18, 381-397.

[49] CANNON, J.R. (1963): Determination of an unknown coefficient in a parabolic differential equation. Duke Math. Journal 30, 313-323.

[50] CANNON, J.R. and DU CHATEAU, P.C. (1978): Determination of unknown coefficients in parabolic operators from overspecified initial-boundary data. Journal of Heat Transfer 100, 503-507.

[51] CANNON, J.R. and DU CHATEAU, P.C. (1980): An inverse problem for a nonlinear diffusion equation. SIAM J. Appl. Math. 39, 272-289.

[52] CANNON, J.R. and DU CHATEAU, P.C. (1980): An inverse problem for an unknown source term in a heat equation. Journal Math. Anal. Appl. 75, 465-485.

[53] CANNON, J.R. and DU CHATEAU, P.C. (1983): An inverse problem for an unknown source term in wave equation. SIAM J. Appl. Math. 43, 553-564.

[54] CANNON, J.R. and EWING, R.E. (1979): Quasilinear parabolic systems with nonlinear boundary conditions. In: [C13], 35-43.

[55] CANNON, J.R. and ZACHMANN, D. (1982): Parameter determination in parabolic partial differential equations from overspecified boundary data. Int. J. Engng. Sci. 20, 779-788.

[56] CARASSO, A.S. (1982): Determining surface temperatures from interior observations. SIAM J. Appl. Math. 42, 558-574.

[57] CARASSO, A.S. (1983): A stable marching scheme for an ill-posed initial value problem. In [C28], 11-35.

[58] CARSLAW, H.S. and JAEGER, J.C. (1959): Conduction of heat in solids. Oxford at the Clarendon Press.

[59] CHADAN, K. and SABATIER, P.C. (1977): Inverse problems in quantum scattering theory. Berlin-Heidelberg-New York, Springer.

[60] CHAMBERS, J.M. (1973): Fitting nonlinear models- numerical techniques. Biometrika 60, 1-13.

[61] CHANDA, K. (1976): Efficiency and robustness of least squares estimators. Sankhya 38, Series B, 153-163.

[62] CHAVENT, G. (1973): Identification of distributed parameters. In: Identification and systems parameter estimation (ed. P.Eykhoff). Proc. 3rd IFAC Symp., The Hague, Delft, 649-660.

[63] CHAVENT, G. (1979): About the stability of the optimal control solution of inverse problems. In: [C13], 45-58.

[64] CHAVENT, G. (1982): Local stability of the output least square parameter estimation technique. INRIA Report no. 136, Le Chesnay Cedex, France.

[65] CHAVENT, G. (1983): Stability of parameters estimated by the output least square method. In: [C28], 37-45.

[66] CHAVENT, G. (1983): Stabilité des paramètres estimés par la méthode des moindres carrés. Revue du CETHEDEC 20, no. 77.

[67] CHEN, W.H. and SEINFELD, J.H. (1972): Estimation of spatially varying parameters in partial differential equations. Int. J. Control 15, 487-495.

[68] CHEN, W.H. and SEINFELD, J.H. (1975): Optimal location of process measurements. Int. J. Control 21, 1003-1014.

[69] CLARK, R.M. (1977): Non-parametric estimation of a smooth regression function. J.Roy.Stat.Soc. B39, 107-113.

[70] COLLATZ, L. (1968): Funktionalanalysis und numerische Mathematik. Springer, Berlin-Heidelberg-New York.

[71] COLLINS, P.L. and KHATRI, H.C. (1969): Identification of distributed parameter systems using finite differences. Trans. ASME Journal of Basic Eng., Series D, 91, 239-245.

[72] COLTON, D. (1979): The approximation of solutions of the backwards heat equation in a nonhomogeneous medium. J. Math. Anal. and Appl. 72, 418-429.

[73] COLTON, D. (1979): The approximation of solutions to the backwards heat equation by solutions of pseudoparabolic equations. In: [C13], 59-71.

[74] COLTON, D. and REEMTSEN, R. (1983): A numerical method for solving the inverse Stefan problem in two space variables. In: [C28], 57-63.

[75] CORMACK, A.M. (1981): Early tomography and related topics. In: [C19], 1-6.

[76] CRANK, J. (1956): The mathematics of diffusion. Oxford at the Clarendon Press.

[77] CRAVEN, P. and WAHBA, G. (1979): Smoothing noisy data with spline functions. Numer. Mathematik 31, 377-403.

[78] CULLUM, J. (1971): Numerical differentiation and regularization. SIAM J. Numer. Anal. 8, 254-265.

[79] CULLUM, J. (1979): The effective choice of the smoothing norm in regularization. Math. Computation 33, 149-170.

[80] CURTAIN, R.F. and ICHIKAWA, A. (1978): Optimal location of sensors for filtering for distributed systems. In: [C11], 236-255.

[81] DAMERT, K., BALZER, D. and REINIG, G. (1976): Nichtlineare Optimierung für Modellierung und Prozeßsteuerung. Akademie-Verlag Berlin.

[82] DANILIN, A.R. and TANANA, V.P. (1984): Necessary and sufficient conditions for the convergence of approximations to linear ill-posed problems in the Hilbert space. Shurnal Vych. Mat. i Mat. Fiz. 24, 633-639.

[83] DAVIES, A.R. (1982): On the maximum likelihood regularization of Fredholm convolution equations of the first kind. In [C22], 95-105.

[84] DAVIES, A.R. et al. (1983): A comparison of statistical regularization and Fourier extrapolation methods for numerical deconvolution. In: [C26], 320-334.

[85] DE HOOG, F.R. (1980): Review of Fredholm equations of the first kind. In: [C17], 119-134.

[86] DE HOOG, F. and WEISS, R. (1973): High order methods for first kind Volterra equations. SIAM J. Numer. Anal. 10, 647-664.

[87] DENNIS, J.E. (1973): Some computational techniques for the nonlinear least squares problem. In: Numerical treatment of systems of nonlinear algebraic equations (eds. G.D.Byrne, C.A.Hall). Academic Press, New York-London, 157-183.

[88] DEUFELHARD, P. (1974): A modified Newton-method for the solution of ill-conditioned systems of nonlinear equations with applications to multiple shooting. Numer. Math. 22, 289-315.

[89] DE VILLERS, N. and GLASSER, D. (1981): A continuation method for nonlinear regression. SIAM J. Numer. Anal. 18, 1139-1154.

[90] DIETZE, S. (1981): Steuer-Approximationsprobleme und Regularisierungsverfahren. Preprints 07-06-81 - 07-09-81, Techn. Univer. Dresden, Sektion Mathematik.

[91] DIEUDONNÉ, J. (1971): Grundzüge der modernen Analysis. Dt. Verlag d. Wissenschaften, Berlin. Aus dem Englischen.

[92] DOLGOPOLOVA, T.F. and IVANOV, V.K. (1966): On numerical differentiation. Shurnal Vych. Mat. i Mat. Fiz 6, 570-576. In Russian.

[93] DOUGLAS, J. and JONES, B.F. (1962): Determination of a coefficient in a parabolic equation. Part II. J. Math. Mech. 11, 919-926.

[94] DRAPER, N.R. and NOSTRAND, R.C. (1979): Ridge regression and James Stein estimation: Review and comments. Technometrics 21, 451-466.

[95] DÜMMEL, S. (1981): Ein inverses Problem für die Wärmeleitungsgleichung. Wiss. Z. d. Techn. Hochschule Karl-Marx-Stadt 23, 355-359.

[96] DÜMMEL, S. (1984): Inverse problems for the heat equation. In: [C30], 21-26.

[97] DUSEMOND, D. and RÖSLER, R. (1977): Ein schneller Algorithmus zur Berechnung von Filteroperationen für gravimetrische Messungen. Gerlands Beitr. Geophysik 86, 165-170.

[98] DYMKE, N. and HOFMANN, B. (1982): Ein Verfahren zur Entfaltung hochenergetischer Szintillations-Gammaspektren bis 8 MeV. Staatl. Amt f. Atomsicherheit und Strahlenschutz der DDR. Report SAAS-293.

[99] ECKHARDT, U. (1976): Zur numerischen Behandlung inkorrekt gestellter Aufgaben. Computing 17, 193-206.

[100] ELDÉN, L. (1977): Algorithms for the regularization of ill-conditioned least-squares problems. BIT 17, 134-145.

[101] ELDÉN, L. (1979): Regularization of the backwards solution of parabolic problems. In: [C13], 73-81.

[102] ELDÉN, L. (1979): A program for interactive regularization. Part I: Numerical algorithms. Report LiTH-MAT-R-79-25, Linköping Univ. Sweden.

[103] ELDÉN, L. (1982): A weighted pseudoinverse, generalized singular values, and constrained least squares problems. BIT 22, 487-502.

[104] ELDÉN, L. (1982): Time discretization in the backward solution of parabolic equations. Math. Comp. 39, 53-84.

[105] ELDÉN, L. (1984): An algorithm for the regularization of ill-conditioned banded least squares problems. SIAM J. Sci. Stat. Comput. 5, 237-254.

[106] ELDÉN, L. (1984): An efficient algorithm for the regularization of ill-conditioned least squares problems with triangular Toeplitz matrix. SIAM J. Sci. Stat. Comput. 5, 229-236.

[107] ELFVING, T. (1978): A method for computing the maximum entropy solution of a linear system. Preprint LiTH-MAT-R-1978-4, Linköping Univ., Dept. of Mathematics, Sweden.

[108] ENGL, H.W. (1980): Analysis und Numerik schlecht gestellter Probleme. Vorlesungsskripte. Joh.-Kepler-Univ. Linz, Austria.

[109] ENGL, H.W. (1981): On least-squares collocation for solving linear integral equations of the first kind with noisy right-hand side. Report no. 199, Inst. f. Math, Joh.-Kepler-Univ. Linz, Austria.

[110] ENGL, H.W. (1981): Stability of solutions to linear operator equations of the first kind and second kind under perturbation of the operator with rank change. Lecture Notes in Med. Inf. 8, Springer, Berlin-New York, 13-28.

[111] ENGL, H.W. (1981): Necessary and sufficient conditions for convergence of regularization methods for solving linear operator equations of the first kind. Numer. Funct. Anal. and Optim. 3, 201-222.

[112] ENGL, H.W. (1983): Regularization by least-squares collocation. In [C26], 345-354.

[113] ENGL, H.W. (1983): On the convergence of regularization methods for ill-posed linear operator equations. In [C28], 81-95.

[114] ENGL, H.W. (1984): On the choice of the regularization parameter for iterated Tikhonov-regularization of ill-posed problems. Report no. 267, Inst. f. Math., Joh.-Kepler-Univ. Linz, Austria.

[115] ENGL, H.W. and NASHED, M.Z. (1981): New extremal characterizations of generalized inverses of linear operators. Journal Math. Anal. Appl. 82, 566-586.

[116] ENGL, H.W. and NEUBAUER, A. (1984): A variant of Marti's method for solving ill-posed linear integral equations that leads to optimal convergence rates. Report no. 266, Inst. f. Math., Joh.-Kepler-Univ. Linz, Austria.

[117] ENGL, H.W. and NEUBAUER, A. (1985): Optimal discrepancy principles for the Tikhonov regularization of integral equations. In: Constructive methods for the practical treatment of integral equations (eds. G.Hämmerlin, K.-H. Hoffmann). Birkhäuser, Basel.

[118] ENGLAND, R. (1983): Some examples of parameter estimation by multiple shooting. In: [C26], 122-136.

[119] ERIKSSON, G. and DAHLQUIST, G. (1983): On an inverse non-linear diffusion problem. In: [C26], 238-245.

[120] EVANS, I.G. (1965): Bayesian estimation of parameters of a multivariate normal distribution. J.Roy.Stat.Soc. B27, 279-283.

[121] EWING, R.E. (1983): Determination of coefficients in reservoir simulation. In [C26], 206-226.

[122] FADDEEV, A.K., KUBLANOVSKAYA, V.N. and FADDEEVA, V.N. (1968): On the solution of linear algebraic systems with rectangle matrices. Trudy mat. in.-ta V.A. Steklova 96, 76-92. Russian.

[123] FALK, R.S. (1980): Error estimates for the approximate identification of a constant coefficient from boundary flux data. Numer. Funct. Anal. and Optimiz. 2, 121-153.

[124] FALK, R.S. (1983): Error estimates for the numerical identification of a variable coefficient. Math.Comp. 40, 537-546.

[125] FALK, R.S. (1983): Numerical identification of a variable coefficient. In [C27] , 31-42.

[126] FARZAN, R. (1984): Mathematical models of electrodynamic sounding with local anomalies. In: [C30], 53-60. In Russian.

[127] FEDOTOV, A.M. (1978): An estimate of the accuracy of approximate solutions of operator equations of the first kind with random errors in the initial data. Soviet Math. Dokl. 19, 652-655.

[128] FEDOTOV, A.M. (1982): Linear ill-posed problems with random errors in the data. Nauka, Sib. otd., Novosibirsk.

[129] FICHTENHOLZ, G.M. (1964): Differential- und Integralrechnung. VEB Dt. Verlag d. Wissenschaften, Berlin. Übers. a.d. Russisch.

[130] FRANKLIN, J.N. (1970): Well-posed stochastic extensions of ill-posed linear problems. J.Math.Anal.Appl. 31, 682-715.

[131] FRANKLIN, J.N. (1974): On Tikhonov's method for ill-posed problems. Math.Comp. 28, 889-907.

[132] FRANZONE, P.C. (1983): Some inverse problems in electrocardiology. In: [C26], 180-205.

[133] FRIEDRICH, C. and HOFMANN, B. (1983): Nichtkorrekte Aufgaben in der Rheometrie. Rheologica Acta 22, 425-434.

[134] FRIEDRICH, V. (1975): Zur iterativen Behandlung unterbestimmter und nichtkorrekter linearer Aufgaben. Beitr. z. Numer. Math 3, 11-20.

[135] FRIEDRICH, V. (1977): Regularisatoren zur numerischen Auswertung indirekter Messungen. Wiss. Zeitschr. d. Techn. Hochschule Karl-Marx-Stadt, 19, 433-436.

[136] FRIEDRICH, V. (1979): Zur stochastischen Regularisierung nichtkorrekter Gleichungen in Hilberträumen. In [C13 ] , 83-88.

[137] FRIEDRICH, V. and HOFMANN, B. (1984): A predictor-corrector technique for constrained least-squares regularization. Wiss. Information d. Sektion Mathematik d. Techn. Hochschule Karl-Marx-Stadt 46.

[138] FRIEDRICH, V., HOFMANN, B. and TAUTENHAHN, U. (1979): Möglichkeiten der Regularisierung bei der Auswertung von Meßdaten. Wiss. Schriftenreihe der Techn. Hochschule Karl-Marx-Stadt 10.

[139] FRIEDRICH, V. and UHLIG, A. (1979): Zur stochastischen Regularisierung linearer Gleichungen in Hilberträumen. Beitr. z. Num. Mathematik 7, 33-48.

[140] FRIND, E.O. and PINDER, G.F. (1973): Galerkin solution of the inverse problem for aquifer transmissivity. Water Res. Res. 9, 1397-1410.

[141] FURI, M. and VIGNOLI, A. (1969): On the regularization of a nonlinear ill-posed problem in Banach spaces. J.Optim. Theory and Appl. 4, 206-209.

[142] GALLIGANI, I. (1979): Quasi-linearization and the identification of parameters in partial differential equations of parabolic type. In [C13], 89-98.

[143] GANDER, W. (1981): Least squares with a quadratic constraint. Numer.Math. 36, 291-307.

[144] GAPONENKO, Y.L. (1981): The method of contracting compact sets for the solution of nonlinear ill-posed problems. Shurnal Vych. Mat. i Mat. Fiz. 21, 1365-1375. In Russian.

[145] GAVURIN, M.K. (1971): Lectures on numerical methods. Nauka, Moscow. In Russian.

[146] GIKHMAN, I.I. and SKOROKHOD, A.V. (1971): Theory of random processes. Nauka, Moscow. In Russian.

[147] GILYAZOV, S.F. (1983): Regularizing subgradient methods for the minimization of convex functions. In [C25], 71-79. In Russian.

[148] GILYAZOV, S.F. and KIRSANOVA, N.N. (1982): On a route for the numerical solution of integral equations of the first kind of Fredholm type in the class of smooth functions. In [C24], 34-47. In Russian.

[149] GILL, P.E., MURRAY, W. and WRIGHT, M.H. (1981): Practical optimization. Academic Press, London.

[150] GLASHOFF, K. (1979): Über die Behandlung inverser Probleme bei streng zeichenfesten Kernen. In: [C13], 99-104.

[151] GLASKO, V.B. (1984): Inverse problems in mathematical physics. Izd. Mosk. Univ., Moscow. In Russian.

[152] GLASKO, V.B. et al. (1979): On an inverse problem of heat conduction. Shurnal Vych. Mat. i Mat. Fiz. 19, 768-774. In Russian.

[153] GLASKO, V.B. et al. (1984): Some problems of heating process optimization. Shurnal Vych. Mat. i Mat. Fiz. 24, 686-693. In Russian.

[154] GOLDMAN, N.L. (1978): The approximate solution of integral equations of the first kind of Fredholm type in the class of piecewise convex functions with bounded first derivative. In: Numer. analysis in FORTRAN. Izd. Mosk. Univ., Moscow, 30-40. In Russian.

[155] GOLDMAN, N.L. (1982): On a modification to the method of descriptive regularization for the solution of Fredholm integral equations of the first kind. In [C24], 48-64. In Russian.

[156] GOLDMAN, N.L. (1983): Numerical solution of nonlinear inverse problems in heat conduction based on the method of descriptive regularization. In: [C29], 8-26. In Russian.

[157] GOLDMAN, N.L. (1983): On an approximate method for the solution of nonlinear inverse problems in multidimensional parabolic equations. In [C29], 69-81. In Russian.

[158] GOLDMAN, N.L. and MOROZOV, V.A. and SAMARIN, M.K. (1977): The method of descriptive regularization and the quality of the approximate solutions. Inshen. Fiz. Shurnal 33, 1117-1124. In Russian.

[159] GOLUB, G.H., HEATH, M. and WAHBA, G. (1979): Generalized cross-validation as a method for choosing a good ridge parameter. Technometrics 21, 215-223.

[160] GOLUB, G.H. and REINSCH, C. (1970): Singular value decomposition and least square solutions. Numer. Math. 14, 403-420.

[161] GONCHARSKI, A.V., LEONOV, A.S. and JAGOLA, A.G. (1972): On a regularization algorithm for ill-posed problems with an approximately given operator. Shurnal Vych. Mat. i Mat. Fiz. 12, 1592-1594. In Russian.

[162] GONCHARSKI, A.V., LEONOV, A.S. and JAGOLA, A.G. (1973): The generalized discrepancy principle. Shurnal Vych. Mat. i Mat. Fiz. 13, 294-302. In Russian.

[163] GONCHARSKI, A.V., LEONOV, A.S. and JAGOLA, A.G. (1975): On the applicability of the discrepancy principle in the case of nonlinear ill-posed problems. Shurnal Vych. Mat. i Mat. Fiz 15, 290-297. In Russian.

[164] GONCHARSKI, A.V. and STEPANOV, V.V. (1979): Algorithms for the approximate solution of ill-posed problems on compact sets. Dokl. Akad. Nauk SSSR, 245, 1296-1299. In Russian.

[165] GONCHARSKI, A.V. and JAGOLA, A.G. (1969): On uniform approximation of monotonous solutions of ill-posed problems. Dokl. Akad. Nauk SSSR 184, 771-773. In Russian.

[166] GOODSON, R.E. and POLIS, M.P. (1978): Identification of parameters in distributed systems. In: Distributed parameter systems-Identification, estimation and control (ed. W.H.Ray, D.G. Lainiotis), Marcel Dekker Inc, New-York-Basel, 47-133.

[167] GORBUNOV, V.K. (1983): Stable reduction of linear integral equations of the first kind. Preprint Akad. Nauk Kirgis. SSR, Frunse. In Russian.

[168] GORBUNOV, V.K. (1984): Methods of reduction for instable problems in numerical mathematics. Izd.-vo ILIM, Frunse. Russian.

[169] GORDONOVA, V.I. and MOROZOV, V.A. (1973): Numerical algorithms for the parameter choice in regularization. Shurnal Vych. Mat. i Mat. Fiz. 13, 539-545. In Russian.

[170] GORENFLO, R. (1979): Numerical treatment of Abel-integral equations. In [C13], 125-133.

[171] GREBENNIKOV, A.I. (1983): On the optimization of solution meth. for some classes of ill-posed problems. Numerical methods and Programming 39, Izd.-vo Mosk. Univ., Moscow. In Russian.

[172] GROETSCH, C.W. (1980): On the Kryanev-Lardy method for ill-posed problems. Math. Nachrichten 96, 27-31.

[173] GROETSCH, C.W. (1980): On a class of regularization methods. Boll. Un. Mat. Ital., Series 17-B, 1411-1419.

[174] GROETSCH, C.W. (1982): On a regularization-Ritz method for Fredholm equations of the first kind. J. Integral Equations 4, 173-182.

[175] GROETSCH, C.W. (1983): On the asymptotic order of accuracy of Tikhonov regularization. J. Optimiz. Th. Appl. 41, 293-298.

[176] GROETSCH, C.W. (1983): Comments on Morozov`s discrepancy principle. In [C28], 97-104.

[177] GROETSCH, C.W. (1984): The theory of Tikhonov regularization for Fredholm equations of the first kind. Pitman Publ. Ltd, Boston.

[178] GROETSCH, C.W. and GUACANEME, J. (1985): Regularized Ritz approximations for Fredholm equations of the first kind. Preprint Series Problemi non ben posti ed inversi, Consiglio Nazionale delle Ricerche, Florence, Italy.

[179] GROETSCH, C.W. and KING, J.T. (1983): The saturation phenomena for Tikhonov regularization. J. Austral. Math. Soc. Ser. A 35, 254-262.

[180] GROETSCH, C.W., KING, J.T. and MURIO, D. (1982): Asymptotic analysis of a finite element method for Fredholm equations of the first kind. In [C22] , 1-12.

[181] GROETSCH, C.W. and SCHOCK, E. (1985): Asymptotic convergence rate of Arcangeli's method for ill-posed problems. Appl. Anal.

[182] GROSSMANN, C. and KAPLAN, A.A. (1979): Strafmethoden und modifizierte Lagrangefunktionen in der nichtlinearen Optimierung. Teubner, Leipzig.

[183] GROSSMANN, C. and KLEINMICHEL, H. (1976): Verfahren der nichtlinearen Optimierung. Teubner, Leipzig.

[184] HADAMARD, J. (1932): Le Probleme de Cauchy et les equations aux dirivees partielles lineaires hyperboliques. Hermann, Paris.

[185] HÄFNER, F. and GIESEL, R.-J. (1978): Numerical results of parameter identification in partial differential equations. In: Theory of nonlinear operators. Proc. Summer School Berlin 1977, Akademie-Verlag, Berlin, 315-321.

[186] HANSON, R.J. (1971): A numerical method for solving Fredholm integral equations of the first kind using singular values. SIAM J. Numer. Anal. 8, 616-622.

[187] HANSON, R.J. and PHILLIPS, J.L. (1975): An adaptive numerical method for solving linear Fredholm integral equations of the first kind. Numer. Math. 24, 291-307.

[188] HARTLEY, H.O. (1961): The modified Gauss-Newton method for the fitting of non-linear regression functions by least squares. Technometrics 3, 269-280.

[189] HÄUSSLER, W.M. (1983): A local convergence analysis for the Gauss Newton method and Levenberg-Morrison-Marquardt algorithms. Computing 31, 231-244.

[190] HINNEBERG, H.-J., HOFMANN, B. and KAISER, T. (1986): Determination of mobility and concentration profiles in semiconductor films by nonlinear identification in a system of Fredholm integral equations of the first kind. Wiss. Z. TH Karl-Marx-Stadt 28, no.2.

[191] HOERL, A.E. and KENNARD, R.W. (1970): Ridge regression: Biased estimation for nonorthogonal problems. Technometrics 12, 55-67. Applications to nonorthogonal problems. Technometrics 12, 68-82.

[192] HOERL, A.E. and KENNARD, R.W. (1976): Ridge regression. Iterative estimation of the biasing parameter. Comm. in Statistics A5, 77-88.

[193] HOFFMANN, K.-H. (1981): Identifizierungsprobleme bei partiellen Differentialgleichungen. In: Numerische Behandlung von Differentialgleichungen, Vol. 3, Birkhäuser, Basel, 97-116.

[194] HOFFMANN, K.-H. and KORNSTAEDT, H.-J. (1980): Zum inversen Stefan-Problem. In: Numerical treatment of integral equations (eds. J. Albrecht, L. Collatz), Birkhäuser, Basel, 115-143.

[195] HOFFMANN, K.-H. and KORNSTAEDT, H.-J. (1982): Ein numerisches Verfahren zur Lösung eines Identifizierungsproblems bei der Wärmeleitung. In: Numerical treatment of free boundary value problems, Birkhäuser, Basel, 108-126.

[196] HOFFMANN, K.-H. and SPREKELS, J. (1983): On the identification of coefficients of elliptic problems by asymptotic regularization. Preprint no. 17, Univ. Augsburg (FRG), Math. Institut.

[197] HOFFMANN, K.-H. and SPREKELS, J. (1984): On the identification of parameters in general variational inequalities by asymptotic regularization. Preprint no. 24, Univ. Augsburg(FRG)

[198] HOFMANN, B. (1976): Untersuchungen zum Vergleich von Regularisierungsverfahren und zu Strategien bei der Wahl des Regularisierungsparameters. Diplomarbeit, Techn. Hochschule Karl-Marx-Stadt, Sektion Mathematik.

[199] HOFMANN, B. (1979): Über Quelldarstellungen bei einigen linearen Regularisierungsverfahren. Beitr. z. Numer. Math. 7, 75-81.

[200] HOFMANN, B. (1979): Bewertung von Schätzungen bei einem linearen Modell auf der Grundlage des Fiduzialkonzepts. Math. Operationsforsch. u. Statistik, Series Statistics 10, 37-46.

[201] HOFMANN, B. (1979): Statistische und numerische Untersuchungen zum Einfluß empirischer Verteilungskenntnisse auf die stochastische Regularisierung. Dissertation, Techn. Hochschule Karl-Marx-Stadt.

[202] HOFMANN, B. (1980): Eine statistische Methode zur Verbesserung empirisch gewonnener klimatologischer Kenngrößen. Zeitschrift für Meteorologie der DDR, 30, 90-99.

[203] HOFMANN, B. (1981): Die Methode der Quasiinversion als Regularisierungsverfahren im Hilbertraum. Beitr. z. Numer. Math. 9, 105-112.

[204] HOFMANN, B. (1981): Dokumentation und theoretische Grundlagen zum Programm RESREG - Regularisierte Lösung linearer Gleichungssysteme unter Nebenbedingungen. Wiss. Inf. no 26, Techn. Hoch. Karl-Marx-Stadt, Sektion Mathematik.

[205] HOFMANN, B. (1982): Zur Regularisierung nichtlinearer inverser Aufgaben. Wiss. Z. d. Techn. Hochschule Karl-Marx-Stadt 24, 315-323.

[206] HOFMANN, B. (1982): Zur regularisierten Lösung einer nichtlinearen nichtkorrekten Faltungsgleichung. Wiss. Z. d. Päd. Hochsch. Liselotte Herrmann Güstrow 20, 163-178.

[207] HOFMANN, B. (1983): Zur Regularisierung nichtlinearer inverser Aufgaben II. Wiss. Z. d. Techn. Hochschule Karl-Marx-Stadt 25, 334-342.

[208] HOFMANN, B. (1984): On empirical stochastic regularization. In: Computational Mathematics. Banach Center Publications 13, PWN-Polish Scientific Publ., Warsaw, 313-317.

[209] HOFMANN, B. (1983): Die Methode der Regularisierung unter Nebenbedingungen- ein einheitlicher Zugang zur numerischen Lösung inverser Aufgaben. Dissertation B. Techn. Hochsch. Karl-Marx-Stadt.

[210] HOFMANN, B. (1983): Zur Methode der emprischen stochastischen Regularisierung. Beitr. z. Numer. Math. 11, 55-67.

[211] HOFMANN, B. (1984): Ein numerisches Verfahren für die Kleinste-Quadrate-Regularisierung- Theorie und Algorithmus. Wiss. Z. d. Techn. Hochsch. Karl-Marx-Stadt 26, 251-259.

[212] HOFMANN, B. (1984): On the generalized discrepancy principle in regularization. Wiss. Inf. no. 48, Techn. Hochschule Karl-Marx-Stadt, Sektion Mathematik.

[213] HOFMANN, B. and FRIEDRICH, V. (1986): On regularization and discretization control for the numerical solution of inverse problems in parabolic equations. BIT 26.

[214] HOFMANN, B. and LINKE, H.-P. (1981): Zur numerischen Behandlung inverser Probleme der Wärmeleitung. INFO 81. Proc. Conf. Neubrandenburg 1981. ZfR-Inf. 80.14, AdW d. DDR, Zentr. f. Rech. Berlin-Adlershof, 99-105.

[215] HSIAO, G.C. (1983): The finite element method for a class of improperly posed integral equations. In [C28] , 117-131.

[216] HUA, T.A. and GUNST, R.F. (1983): A note on negative ridge parameters. Comm. in Statistics 12, 37-45.

[217] HUMAK, K.M.S. (1977): Statistische Methoden der Modellbildung. Statistische Inferenz für lineare Parameter. Akademie-Verlag, Berlin.

[218] HUMAK, K.M.S. (1983): Statistische Methoden der Modellbildung. Nichtlineare Regression u.a. Probleme. Akademie-Verlag, Berlin.

[219] ISAKOV, V.M. (1979): On uniqueness of solutions for inverse problems of potential theory. In [C13], 135-140.

[220] ISKENDEROV, A.D. (1975): Multidimensional inverse problems for linear and quasilinear parabolic equations. Dokl. Akad. Nauk SSSR 225, 1005-1008. In Russian.

[221] IVANICKIY, A. Y. (1983): On the solution of systems of linear algebraic equations with imprecise data. In [C25], 136-141. In Russian.

[222] IVANOV, V.K. (1962): On linear ill-posed problems. Dokl. Akad. Nauk SSSR 145, 270-272. In Russian.

[223] IVANOV, V.K. (1963): On some ill-posed problems. Mat. Sbornik 61, 211-223. In Russian.

[224] IVANOV, V.K. (1966): On the approximate solution of operator equations of the first kind. Shurnal Vych. Mat. i Mat. Fiz 6, 1089-1094. In Russian.

[225] IVANOV, V.K., VASIN, V.V. and TANANA, V.P. (1978): Theory of linear ill-posed problems and its applications. Nauka, Moscow.

[226] JAGOLA, A.G. (1979): On the parameter choice in regularization by the generalized discrepancy principle. Dokl. Akad. Nauk SSSR 245, 37-39. In Russian.

[227] JAGOLA, A.G. (1980): On the solution of nonlinear ill-posed problems by means of the generalized discrepancy principle. Shurnal Vych. Mat. i Mat. Fiz. 252, 810-813.

[228] JAKEMAN, A.J. and YOUNG, P. (1980): Systems identification and estimation for convolution integral equations. In [C17], 235-255.

[229] JANOVSKAJA, T.B. and POROKHOVA, L.N. (1983): Inverse problems in geophysics. Izd.-vo Leningradsk. Univ., Leningrad.

[230] JANOVSKÝ, V., MAREK, I. and NEUBERG, J. (1983): Maxwell's equations with incident waves as a field source. In [C28], 133-148.

[231] JOCHUM, P. (1982): To the numerical solution of an inverse Stefan problem in two space variables. In: Numerical treatment of free boundary problems. Birkhäuser, Basel, 127-136.

[232] JOHN, F. (1960): Continuous dependence on data for solutions of partial differential equations with a prescribed bound. Comm. on Pure and Appl. Math. 13, 551-585.

[233] JONES, B.F. (1962): Determination of a coefficient in a parabolic equation. Part I. Existence and uniqueness. J. Math. Mech. 11, 907-918.

[234] KAHANE, C.S. (1983): The solution of a mildly singular integral equation of the first kind on a ball. Int. Equ. and Oper. Th. 6, 67-133.

[235] KAISER, T. (1985): Das inverse Problem der Bestimmung von Wärmeübergangsfunktionen bei der Abkühlung von Draht in Kühlstrecken. Diplomarbeit. Techn. Hochschule Karl-Marx-Stadt, Sektion Mathematik.

[236] KANTOROWITSCH, L.W. and AKILOW, G.P. (1964): Funktionalanalysis in normierten Räumen. Akademie-Verlag, Berlin. Aus dem Russischen.

[237] KHJAMARIK, U.A. (1983): Projection methods for the regularization of linear ill-posed problems. Trudy VZ Tartuski Gos. Univ. 50, Tartu, 69-90.

[238] KING, J.T. and CHILLINGWORTH, D. (1979): Approximation of generalized inverses by iterated regularization. Numer. Funct. Anal. and Optimiz. 1, 499-513.

[239] KITAMURA, S. and NAKAGIRI, S. (1977): Identifiability of spatially varying and constant parameters in distributed systems of parabolic type. SIAM J. Control and Optimization 15, 785-802.

[240] KLEIN, G. (1979): On spline functions and statistical regularization of ill-posed problems. J. Computational and Appl. Math. 5, 259-263.

[241] KLIMOV, G.P. (1973): Invariant reasoning in statistics. Izd.-vo Mosk. Univ, Moscow.

[242] KLUGE, R. (1979): Nichtlineare Variationsungleichungen. VEB Dt. Verl. d. Wissenschaften, Berlin.

[243] KLUGE, R. (1983): Zur "Koeffizienten"bestimmung in linearen Operator- und Evolutionsgleichungen. Math. Nachrichten 112, 153-175.

[244] KLUGE, R. and LANGMACH, H. (1978): On the determination of some rheologic properties of mechanical media. In: Theory of nonlin. operators. Proc. Summer School Berlin 1977. Akad.-Verlag, Berlin, 141-158.

[245] KLUGE, R. and LANGMACH, H. (1978): On some problem of determination of functional parameters in pde.s. In [C11] , 289-309.

[246] KNABNER, P. (1983): Regularizing the Cauchy problem for the heat equation by sign restrictions. In [C28], 165-177.

[247] KNOPS, R.J. and PAYNE, L.E. (1979): Uniqueness and continuous dependence of the null-solution in the Cauchy problem for a nonlinear elliptic system. In [C13], 151-160.

[248] KOCOTOV, I.I. (1976): On a new method for choosing the regularization parameter. Shurnal Vych. Mat. i Mat. Fiz. 16, 499-503. In Russian.

[249] KOLMOGOROV, A.N. and FOMIN, S.V. (1975): Reelle Funktionen und Funktionalanalysis. VEB Dt. Verl. d. Wissenschaften, Berlin. Aus dem Russischen.

[250] KONDRATEV, K.Y. and TIMOFEEV, Y.M. (1970): Thermal sounding of the atmosphere by satellites. Gidromet. izd.-vo, Leningrad. In Russian.

[251] KRABS, W. (1982): Optimal control of processes governed by partial differential equations I. Heating processes. Z. f. Operat. Research, Series A 26, 21-48.

[252] KREIN, M. (1954): On an effective solution method for inverse boundary problems. Dokl. Akad. Nauk SSSR 94, 987-990. Russian.

[253] KRISHNAPRASAD, P.S. and BARAKAT, R. (1977): A descent approach to a class of inverse problems. J. Comp. Phys. 24, 339-347.

[254] KRUKOVSKI, P.G. (1980): Numerical methods for the solution of inverse problems in heat conduction and the identification of thermal parameters. In: Systems Analysis and Simulation (ed. A. Sydow), Akademie-Verlag, Berlin, 336-341. In Russian.

[255] KUHNERT, F. (1976): Pseudoinverse Matrizen und die Methode der Regularisierung. Teubner, Leipzig.

[256] KUHNERT, F. (1980): Über numerische Verfahren für inverse Matrizeneigenwertprobleme. Beitr. z. Numer. Math. 8, 75-84.

[257] KUHNERT, F. (1981): Übersicht über numerische Verfahren für das Matrixeigenwertproblem. Berichte der MG DDR- Mathematikerkongr. der DDR 1981. Vortragsauszüge, no. 1, 66-79.

[258] LADYSHENSKAJA, O.A., SOLONNIKOV, V.A. and URALCEVA, N.N. (1967): Linear and quasilinear equations of parabolic type. Nauka,Moscow. In Russian.

[259] LANDWEBER, L. (1951): An iteration formula for Fredholm integral equations of the first kind. Amer. J. Math. 73, 615-624.

[260] LANG, K. (1982): Zur Behandlung inverser Aufgaben für die Wärmeleitungsgleichung unter Berücksichtigung der Fourierschen Methode. Diplomarbeit. Techn. Hochsch. Karl-Marx-Stadt, Sektion Mathematik.

[261] LANGMACH, H. (1979): Zur Bestimmung von Funktionalparametern in partiellen Differentialgleichungen. Dissertation, ZIMM d. AdW. d. DDR, Berlin.

[262] LARKIN, F.M. (1983): The weak Gaussian distribution as a means of localization in Hilbert spaces. In [C27], 145-177.

[263] LATTES, R. and LIONS, J.-L. (1967): Méthode de quasi-réversibilité et applications. Dunod, Paris.

[264] LAVRENTEV, M.M. (1967): Some improperly posed problems in mathematical physics. Springer, Berlin-Heidelberg-New York, Transl. from the Russian.

[265] LAVRENTEV, M.M. (1977): On the regularization of Volterra operator equations. In [C7], 199-205. In Russian.

[266] LAVRENTEV, M.M. (1979): Some problems of analytic continuation. In [C13], 165-170.

[267] LAVRENTEV, M.M. and FEDOTOV, A.M. (1982): On some ill-posed problems of mathematical physics with random input data. Shurnal Vych. Mat. i Mat. Fiz 22, 133-143. In Russian.

[268] LAVRENTEV, M.M. and REZNICKAJA, K.G. and JACHNO, V.G. (1982): Onedimensional inverse problems of mathematical physics. Izd.-vo Nauka, Sib. otd., VZ ANSSSR, Novosibirsk.In Russian.

[269] LAWSON, C.L. and HANSON, R.J. (1974): Solving least squares problems. Prentice-Hall, Englewood Cliffs, New Jersey.

[270] LEONOV, A.S. (1982): On the relations between the generalized discrepancy and the generalized discrepancy principle for the solution of nonlinear ill-posed problems. Shurnal Vych. Mat. i Mat. Fiz. 22, 783-790. In Russian.

[271] LEXIKON Optimierung und Steuerung (1985): Akademie-Verlag, Berlin.

[272] LEXIKON der Stochastik (1975): Akademie-Verl., Berlin. 2.Aufl.

[273] LINDLEY, D.V. and SMITH, A.F. M. (1972): Bayesian estimation for the linear model. J. Roy. Stat. Soc. B 34, 1-18.

[274] LINKE, H.-P. (1979): Ein inverses Problem zur Koeffizientenbestimmung in der Wärmeleitungsgleichung. Wiss. Z. d. Techn. Hochschule Karl-Marx-Stadt 21, 613-617.

[275] LINKE, H.-P. and HOFMANN, B. (1981): Eine inverse dritte Randwertaufgabe für die Wärmeleitungsgleichung. Berichte, Heft 5, Conference IX. IKM Weimar 1981, 93-96.

[276] LINKE, H.-P. and HOFMANN, B. (1982): Zur Bestimmung von Temperaturleitzahlen und Wärmeübergangszahlen bei vorliegender dritter Randwertaufgabe. In [C23], 20-37.

[277] LINZ, P. (1980): A survey of methods for the solution of Volterra integral equations of the first kind. In [C17], 183-194.

[278] LINZ, P. (1982): The solution of Volterra equations of the first kind in the presence of large uncertainties. In [C22], 123-130.

[279] LINZ, P. (1984): Uncertainty in the solution of linear operator equations. BIT 24, 92-101.

[280] LIONS, J.L. (1971): Optimal control of systems governed by partial differential equations. Springer, Berlin-Heidelberg-New York.

[281] LIONS, J.L. (1978): Some aspects of modelling problems in distributed parameter systems. In [C11], 11-41.

[282] LIONS, J.L. and MAGENES, E. (1968): Problème avec limites non homogènes et applications. Vol I. Dunod, Paris.

[283] LISKOVEC, O.A. (1979): Discrete schemes in the regularization method for ill-posed extremum problems. Dokl. Akad. Nauk SSSR 248, 1299-1303. In Russian.

[284] LJUSTERNIK, L.A. and SOBOLEV, V.I. (1968): Elemente der Funktionalanalysis. Akademie-Verlag, Berlin. 4.Aufl. Aus d. Russisch.

[285] LÖCKER, J. and PRENTER, P.M. (1980): Regularization with differential operators. I: J.Math.Anal.Appl. 74, 504-529. II: SIAM J.Numer.Anal 17, 247-267.

[286] LOUIS, A.K. and NATTERER, F. (1983): Mathematical problems of computerized tomography. Proc. IEEE 71, 379-389.

[287] LOTT, W.F. (1973): The optimal set of principal components restrictions on a least-squares regression. Comm. in Statistics 2, 449-464.

[288] LUKAS, M.A. (1980): Regularization. In [C17], 151-182.

[289] MACKENROTH, U. (1981): Optimalitätsbedingungen und Dualität bei zustandsrestringierten parabolischen Kontrollproblemen. Math. Operationsforsch. u. Statistik, Series Optimization 12, 65-89.

[290] MAREK, I. and ŽITNY, K. (1983): Matrix analysis for applied sciences. Teubner, Leipzig.

[291] MARQUARDT, D. (1963): An algorithm for least squares estimation of nonlinear parameters. SIAM J. Appl. Math. 11, 431-441.

[292] MARQUARDT, D. (1970): Generalized inverses, ridge regression, biased linear estimation and nonlinear estimation. Technometrics 12, 591-612.

[293] MARTI, J.T. (1980): On the convergence of an algorithm computing minimum-norm-solutions of ill-posed problems. Math. Comp. 34, 521-527.

[294] MARTI, J.T. (1982): On a regularization method for Fredholm integral equations of the first kind using Sobolev space. In: [C22], 59-66.

[295] MARTI, J.T. (1983): Numerical solution of Fujita's equation. In: [C28] , 179-187.

[296] MC KEOWN, J.J. (1980): Nonlinear least-sqares problems, 91-105, Large residual nonlinear least-squares problems, 107-121, Parametric sensitivity analysis, 387-406. In: Nonlinear optimimization (eds. L.C.W. Dixon et al.), Birkhäuser, Boston.

[297] MEIER, M. (1981): Dokumentation zum Programm WAERME. Wiss. Inf. 23, Techn. Hochschule Karl-Marx-Stadt, Sektion Mathematik.

[298] MENZEL, R., PÖNISCH, G. and SCHWETLICK, H. (1979): Ein Programmpaket zur Lösung nichtlinearer Probleme. Z. f. Rechentechnik/ Datenverarb. 16, 28-29. Dokumentation (1977), TU Dresden.

[299] MEYER, S. (1984): Die Existenz einer Lösung für ein inverses Problem der Wärmeleitungsgleichung im quasilinearen Fall. Wiss. Z. d. Techn. Hochschule Karl-Marx-Stadt 26, 259-264.

[300] MICHEL, B. (1979): About inverse problems in continuum mechanics and solid state physics. In [C13] , 171-180.

[301] MIKHEYEV, M. (1964): Fundamentals of heat transfer. Peace Publ. Moscow. Transl. from the Russian.

[302] MIKHLIN, S.G. (1972): Lehrgang der mathematischen Physik. Akademie-Verlag, Berlin. Übers. a. d. Russischen.

[303] MILLER, G.F. (1974): Fredholm equations of the first kind. In: Numerical solution of integral equations (eds. L.M. Delves, J.Walsh), Clarendon Press, Oxford, 175-188.

[304] MILLER, K. (1970): Least-squares methods for ill-posed problems with a prescribed bound. SIAM J. Math. Anal 1, 52-74.

[305] MILO, W. and WASILEWSKI, Z. (1983): Some remarks on the properties of the chosen Moore-Penrose generalized inverse algorithms. Report. Inst. Econometrics and Statistics. Univ. of Lódź. Poland.

[306] MILSTEIN, J. (1981): The inverse problem: Estimation of kinetic parameters. In: Modelling of chemical reactions (eds. K.H. Ebert et al.). Springer, Berlin-Heidelberg-New York, 92-101.

[307] MITCHELL, T. and DRAPER, N.R. (1982): Using external information in linear regression, with a commentary on ridge regression. Technical Report no. 2327, Madison Res. Center, Univ. of Wiscon.

[308] MIYAMOTO, S., IKEDA, S. and SAWARAGI, Y. (1978): Identification of distributed systems and the theory of the regularization. J. of Math. Anal. and Appl. 63, 77-95.

[309] MORÉ, J.J. (1977): The Levenberg-Marquardt algorithm: Implementation and theory. In: Numerical analysis (ed. G.A.Watson). Springer, Berlin-Heidelberg-New York, 105-116.

[310] MOROZOV, V.A. (1967): On the parameter choice for the solution of functional equations by the regularization method. Dokl. Aka. Nauk SSSR 175, 1225-1228. In Russian.

[311] MOROZOV, V.A. (1968): On the discrepancy principle for the solution of operator equations with the method of regularization. Shurnal Vych. Mat. i Mat. Fiz 8, 295-309. In Russian.

[312] MOROZOV, V.A. (1973): Linear and nonlinear ill-posed problems. Itogi nauki i techniki, mat. analisis, Vol. 11. Izd.-vo VINITI, Moscow, 129-178. In Russian.

[313] MOROZOV, V.A. (1973): On the discrepancy principle in the solution of inconsistent equations by the regularization method of A.N. Tikhonov. Shurnal Vych. Mat. i Mat. Fiz 13, 1099-1111. In Russian.

[314] MOROZOV, V.A. (1979): On the solution of operator equations of the first kind by a finite rank approximation. Dokl. Akad. Nauk SSSR 247, 1317-1320. In Russian.

[315] MOROZOV, V.A. (1979): Regular method for solving nonlinear problems. In: [C13], 289-296.

[316] MOROZOV, V.A. (1979): Some particularities of the numerical sol. of integral equations by the method of descriptive regularization. In: Numerical analysis in FORTRAN, Izd.-vo Mosk. Univ., Moscow, 41-49. In Russian.

[317] MOROZOV, V.A. (1982): On stable numerical methods for the solution of linear algebraic systems of equations. In [C24], 3-10.

[318] MOROZOV, V.A. (1982): The method of quasisolutions on noncompact sets. Soviet Math. Dokl 25, 489-493.

[319] MOROZOV, V.A. (1984): On the stable numerical solution of consistent systems of linear algebraic equations. Shurnal Vych. M. i Mat. Fiz 24, 179-186. In Russian.

[320] MOROZOV, V.A. (1984): Methods for solving incorrectly posed problems. Springer, New York-Berlin-Tokyo. Trans. from Russian.

[321] MOROZOV, V.A. and GILJAZOV, S.F. (1982): Regularization of conditionally well-posed operator equations by the conjugate gradient method. In [C24], 19-28. In Russian.

[322] MOROZOV, V.A. and NAZIMOV, A.B. (1983): On the theory of L-pseudoinverses. In [C25], 20-29. In Russian.

[323] MOROZOV, V.A. and SAMARIN, M.K. (1982): The construction of half bounded monotonic solutions to operator equations. In [C24], 29-33. In Russian.

[324] MOSCO, U. (1969): Convergence of convex sets and solutions of variatonal inequalities. Advances in Mathematics 3, 510-585.

[325] MURAVEVA, M.V. (1973): On optimality and limit properties of Bayesian solutions to linear algebraic systems. Shurnal Vych. Mat. i Mat. Fiz. 13, 819-828. In Russian.

[326] NAKONECNI, A.G. (1979): Minimax estimators in systems with distributed parameters. Preprint 79-46. Akad. Nauk Ukrainsk. SSR, Inst. Kibernetiki, Kiev. In Russian.

[327] NASHED, M.Z. (1974): Approximate regularized solutions to improperly posed linear integral and operator equations. In: Constructive and computational methods for differential and integral equations. Symp. Indiana Univ. 1974, Springer, Berlin- Heidelberg- New York, 289-332.

[328] NASHED, M.Z. (1976): On moment-discretization and least-squares solutions of linear integral equations of the first kind. J. Math. Anal. Appl. 53, 359-366.

[329] NASHED, M.Z. (1981): Continuous and semicontinuous analogues of iterative methods of Cimmino and Kaczmarz with application to the inverse Radon transform. In [C19], 160-178.

[330] NASHED, M.Z. and WAHBA, G. (1974): Convergence rates of approximate least squares solutions of linear integral and operator equations of the first kind. Math. Comp. 28, 69-80.

[331] NATTERER, F. (1977): Regularisierung schlecht gestellter Probleme durch Projektionsverfahren. Numer. Math. 28, 329-341.

[332] NATTERER, F. (1977): The finite element method for ill-posed problems. R.A.I.R.O. Anal. Num. 11, 271-278.

[333] NATTERER, F. (1978): Numerical inversion of the Radon transform. Numer. Math. 30, 81-91.

[334] NATTERER, F. (1979): On the inversion of the attenuated Radon transform. Numer. Math. 32, 431-438.

[335] NATTERER, F. (1980): A Sobolev space analysis of picture reconstruction. SIAM J. Appl. Math. 39, 402-411.

[336] NATTERER, F. (1981): The identification problem in emission computerized tomography. In [C19], 45-56.

[337] NATTERER, F. (1983): Computerized tomography with unknown sources. SIAM J. Appl. Math. 43, 1201-1212.

[338] NATTERER, F. (1983): On the order of regularization methods. In: [C28], 189-203.

[339] NATTERER, F. (1984): Einige Beiträge der Mathematik zur Computer-Tomography. Z. f. Angew. Math. u. Mechanik 64, T252-T260.

[340] NATTERER, F. (1984): Error bounds for Tikhonov regularization in Hilbert scales. Applic. Analysis.

[341] NOCEDAL, J. and OVERTON, L. (1983): Numerical methods for solving inverse eigenvalue problems. Technical Report no. 74, Com. Sc. Dept., New York Univ. Courant Inst of Math. Sc..

[342] NOWAK, U. and DEUFELHARD, P. (1982): Towards parameter identification for large chemical reaction systems. Preprint no. 199, SFB 123, Univ. Heidelberg, Inst. f. Angew. Mathematik.

[343] NOWAK, U. and DEUFELHARD, P. (1983): Numerical identification of selected rate constants in large chemical reaction systems. Preprint no. 228, SFB 123, Univ. Heidelberg, Inst. f. Angew. Mathematik.

[344] NUTBROWN, D.A. (1975): Identification of parameters in linear equations of groundwater flow. Water Res. Res. 11, 581-588.

[345] NYCHKA, D. and WAHBA, G. (1983): Cross-validated spline methods for the estimation of three dimensional tumor size distribution from observations on two dimensional cross sections. Technical Report no. 711, Univ. of Wisconsin, Dept. of Statistics, Madis.

[346] ORTEGA, J.M. and RHEINBOLDT, W.C. (1966): On discretization and differentiation of operators with applications to Newton's method. SIAM J. Numer. Anal. 3, 143-156.

[347] ORTEGA, J.M. and RHEINBOLDT, W.C. (1970): Iterative solution of nonlinear equations in several variables. Academic Press, New York.

[348] OSWALD, P. (1981): $L^p$-Approximation durch Reihen nach dem Haar-Orthogonalsystem und dem Faber-Schauder-System. J.Approx.Th. 33, 1-27.

[349] OVERTON, M.L. (1981): A quadratically convergent method for minimizing a sum of Euclidean norms. Technical Report no. 030, New York Univ., Dept. of Comp. Sc., Courant Inst. of Math. Sc..

[350] PETERS, G. and WILKINSON, J.H. (1979): Inverse iteration, ill-conditioned equations and Newton's method. SIAM Review 21, 339-360.

[351] PETROV, A.P. (1967): Estimations for linear functionals in the solution of ill-posed inverse problems. Shurnal Vych. Mat. i Mat. Fiz. 7, 648-654. In Russian.

[352] PHILLIPS, D.L. (1962): A technique for the numerical solution of certain integral equations of the first kind. JACM 9, 84-97.

[353] POLIS, M.P. and GOODSON, R.E. (1976): Parameter identification in distributed systems. A synthesizing overview. Proc. IEEE 64, 45-61.

[354] POLYAK, B.T. (1964): Gradient methods for the solution of equations and inequalities. Shurnal Vych. Mat. i Mat. Fiz. 4, 995-1006. In Russian.

[355] PRILEPKO, A.I. (1974): Über die Existenz und Eindeutigkeit von Lösungen inverser Probleme. Math. Nachrichten 63, 135-153.

[356] PUKHNACEVA, T.P. (1982): Inverse problems for the Maxwell equations in a medium possessing anisotropic conductivity. Diff. Ur. 18, 1780-1787. In Russian.

[357] PROVENCHER, S.W. and VOGEL, R.H. (1983): Regularization techniques for inverse problems in molecular biology. In [C26], 304-319.

[358] PYTEV, J.P. (1982): The pseudoinverse operator, properties and applications. Mat. Sbornik 118, 19-49.

[359] PYTEV, J.P. (1983): The reduction problem in experimental investigations. Mat. Sbornik 120, 240-272.

[360] RAMSIN, H. and WEDIN, P.-A. (1977): A comparison of some algorithms for the nonlinear least squares problem. BIT 17, 72-90.

[361] RAO, C.R. (1965): The theory of least squares when the parameters are stochastic and its application to the analysis of growth curves. Biometrika 52, 447-458.

[362] RAO, C.R. (1973): Lineare statistische Methoden und ihre Anwendungen. Akademie-Verlag, Berlin. Übers. a. d. Englischen.

[363] RAO, C.R. (1975): Simultaneous estimation of parameters in different linear models and applications to biometric problems. Biometrics 31, 545-554.

[364] RHEINBOLDT, W.C. (1976): On measures of ill-conditioning for nonlinear equations. Math. Comp. 30, 104-111.

[365] RICHTER, C., GROSSMANN, C. et al. (1980): Programmpaket NLOPT-Nichtlineare Optimierung. Dokumentation in Preprints 07-22-80, 07-23-80, 07-24-80 and 07-25-80. Techn. Univ. Dresden, Sektion Mathematik.

[366] RICHTER, G.R. (1978): Numerical solution of integral equations of the first kind with nonsmoothing kernels. SIAM J. Numer. Anal. 15, 511-522.

[367] RICHTER, G.R. (1981): Numerical identification of a spatially varying diffusion coefficient. Math. Comp. 36, 375-386.

[368] RICHTER, J.-U. (1984): Rekonstruktion von Flußparameterfunktionen in einem Modell der Dotantenumverteilung. Diplomarbeit. Tech. Hochschule Karl-Marx-Stadt, Sektion Mathematik.

[369] RICHTER, M. (1979): Berechnung der Verteilungsfunktion der Zufallsgröße $\int_0^T \chi_t^2(\omega)$ dt und deren Anwendungen. Wiss. Z. d. Techn. Hochschule Karl-Marx-Stadt 21, 633-643.

[370] ROMANOV, V.G. (1975): Volterra operator equations of the first kind. Classes of uniqueness. In: Some problems of numerical and applied mathematics. Izd.-vo Nauka, Sib. otd., Novosibirsk, 123-135. In Russian.

[371] ROMANOV, V.G. (1979): Inverse problems and energy inequalities. In [C13], 215-222.

[372] ROMANOV, V.G. (1984): Inverse problems of mathematical physics. Izd.-vo Nauka, Moscow. In Russian.

[373] RÖSLER, R. (1980): Inverse Aufgaben in der Geophysik. In [C12], 4-14.

[374] ROWLAND, S.W. (1979): Computer implementation of image reconstruction formulas. In [C14], 9-79.

[375] RUHE, A. (1979): Accelerated Gauss-Newton algorithms for nonlinear least squares problems. BIT 19, 356-367.

[376] RUTMAN, R.S. and CABRAL, L.M. (1982): Descriptive regularization of the Fredholm integral equations of the first kind. In [C22], 313-323.

[377] SABATIER, P.C. (1977): Positivity constraints in linear inverse problems I. General theory. Geophys. J. R. atr. Soc. 48, 415-441.

[378] SABATIER, P.C. (1983): Critical analysis of the mathematical methods used in electromagnetic inverse theories: a quest for new routes in the space of parameters. Preprint PM/83/8. Univ. des Sci. et Techn. du Languedoc, Montpellier, France.

[379] SAMARIN, M.K. (1976): Least squares approximation of piecewise monotonous functions. In: Numerical analysis in FORTRAN. Univ. Moscow Press, Vyp. 15, 83-90. In Russian.

[380] SAMARIN, M.K. (1980): A uniform approximation of monotonous functions. In: Numerical analysis in FORTRAN. Univ. Moscow Press, 8-14. In Russian.

[381] SAMARSKI, A.A. (1977): Theory of difference schemes. Izd.-vo Nauka, Moscow. In Russian.

[382] SCHITTKOWSKI, K. and STOER, J. (1979): A factorization method for the solution of constrained linear least squares problems allowing subsequent data changes. Numer. Math. 31, 431-463.

[383] SCHMIDT, R. (1982): Advances in nonlinear parameter optimization. Springer, Berlin-Heidelberg-New York.

[384] SCHNEIDER, J. (1980): Zur thermischen Theorie der Verarbeitung und Viskosimetrie von Polymerschmelzen. Plaste und Kautschuk 27, 361-365.

[385] SCHNEIDER, J. (1984): Über einige mechanisch-thermisch gekoppelte Probleme der Fluidrheologie und der Verarbeitungstechnik. In: [C30], 186-197.

[386] SCHOCK, E. (1982): Numerische Lösung Fredholmscher Integralgleichungen. Universität Kaiserslautern.

[387] SCHOCK, E. (1983): On the approximate solution of ill-posed equ. in Banach spaces. In: Proc. Conf. Essen 1982, 351-362. FB Math. Universität Essen.

[388] SCHOCK, E. (1983): Regularisierungsverfahren für Gleichungen erster Art mit positiv definiten Operatoren. In [C28], 227-236.

[389] SCHOCK, E. (1984): On the asymptotic order of accuracy of Tikhonov regularization. J. Optimiz. Theory and Appl. 44.

[390] SCHOCK, E. (1985): Parameter choice by discrepancy principles for Tikhonov regularization of ill-posed problems. Integral Equ. and Operator Theory.

[391] SCHOCK, E. (1985): Ritz-regularization versus least-square regularization. Solution methods for integral equations of the first kind. J. Integral Equations.

[392] SCHWEFEL, H.-P. (1977): Numerische Optimierung von Computer-Modellen mittels der Evolutionsstrategie. Birkhäuser, Basel-Stuttgart.

[393] SCHWETLICK, H. (1973): Zur Konvergenz regularisierter Gauß-Newton-Verfahren. Shurnal Vych. Mat. i Mat. Fiz. 13, 1371-1382.

[394] SCHWETLICK, H. (1979): Numerische Lösung nichtlinearer Gleichungen. VEB Dt. Verl. d. Wissenschaften, Berlin.

[395] SCHWETLICK, H. and TILLER, V. (1982): Numerical methods for estimating parameters in nonlinear models with errors in the variables. Preprint no. 69, Martin-Luther-Univ. Halle-Wittenberg, Sektion Mathematik.

[396] SEIDMAN, T.I. (1979): Ill-posed problems arising in boundary control and observation for diffusion equations. In:[C13], 233-248.

[397] SEIDMAN, T.I. (1980): Nonconvergence results for the application of least-squares estimation to ill-posed problems. J. Optimiz. Th. and Appl. 30, 535-547.

[398] SEIDMAN, T.I. (1981): Convergent approximation methods for ill-posed problems. Control and Cybernetics 10, 31-49 and 51-71.

[399] SEIDMAN, T.I. and ZHOU, H,-X. (1982): Existence and uniqueness of optimal controls for a quasilinear parabolic equation. SIAM J. Control and Optimization 20, 747-762.

[400] SEINFELD, J.H. and KODA, M. (1978): Numerical implementation of distributed parameter filters with application to problems in air pollution. In:[C11], 42-69.

[401] SHUKOVSKI, E.L. and MOROZOV, V.A. (1972): On iterative Bayesian regularization of algebraic equation systems. Shurnal Vych. Mat. i Mat. Fiz. 12, 464-465.

[402] SMIRNOV, V.I. (1962): Lehrgang der Höhrren Mathematik. VEB Dt. Verlag d. Wissenschaften. Übers. a. d. Russischen.

[403] SMITH, T.B. and SHANNO, D.F. (1971): An improved Marquardt procedure for non-linear regressions. Technometrics 13, 63-74.

[404] SONNEVEND, G. (1984): Solution of identification problems via realization theory. In: Mathematical models in physics and chemistry (eds. A.A.Samarski, J.Kátai), Teubner, Leipzig, 107-115.

[405] STADLER, E. (1984): Die Lösung eines inversen Eigenwertproblems mit der Regularisierungsmethode. Wiss. Z. d. Techn. Hochschule Karl-Marx-Stadt 26, 299-305.

[406] STEEN, N.M. and BYRNE, G.D. (1973): The problem of minimizing nonlinear functionals I. Least squares. In: Numerical solution of systems of nonlinear algebraic equations (eds. G.D. Byrne, C.A.Hall), Academic Press, New York-London, 185-239.

[407] STEPANOV, A.A. (1983): On numerical algorithms for the regularization parameter by cross-validation. In:[C25], 126-135. In Russian.

[408] STOYAN, G. (1979): Identification of a spatially varying coefficient in a parabolic equation. In:[C13] , 249-258.

[409] STOYAN, G. (1980): On the identification of diffusion coefficients. Banach Center Publication 3, Warsaw, 367-377.

[410] STOYAN, G. and BAUMERT, H. (1981): Parameter identification in transverse mixing models of rivers- an inverse problem for a parabolic equation. Zeitschr. f. Ang. Math. u. Mech. 61, 617-627.

[411] STRAKHOV, V.N. (1970): On the solution of linear ill-posed problems in the Hilbert space. Different. Uravnenia 6, 1490-1495. In Russian.

[412] STRAKHOV, V.N. and VALYASHKO, G.M. (1981): Algorithms for the adaptive regularization of linear ill-posed problems. Dokl. Aka. Nauk SSSR 259, 546-548. In Russian.

[413] STRAND, O.N. (1974): Theory and methods related to the singular function expansion and Landwebers iteration for integral equations of the first kind. SIAM J. Numer. Anal. 11, 798-825.

[414] STRAND, O.N. and WESTWATER, E.R. (1968): Statistical estimation of the numerical solution of a Fredholm integral equation of the first kind. JACM 15, 100-114.

[415] STRAND, O.N. and WESTWATER, E.R. (1968): Minimum-RMS-estimation of the numerical solution of a Fredholm integral equation of the first kind. SIAM J. Numer. Anal 5, 287-295.

[416] STRAWDERMAN, W.E. (1978): Minimax adaptive generalized ridge regression estimators. J. Amer. Statist. Assoc. 73, 623-627.

[417] SUZUKI, T. (1982): Remarks on the uniqueness in an inverse problem for the heat equation. Proc. of the Japan Academy, Ser. 58 A 93-96 and 175-177.

[418] SUZUKI, T. (1982): Deformation formulas and their applications to spectral and evolutional inverse problems. Lect. Notes in Num. Appl. Anal. 5, 289-311.

[419] SUZUKI, T. (1983): Uniqueness and nonuniqueness in an inverse problem for the parabolic equation. Journal of Differential Equ. 47, 296-316.

[420] SWAMY, P.A.V.B. (1971): Statistical inference in random coefficient regression models. Springer Berlin-Heidelberg-New York.

[421] TAUTENHAHN, U. (1979): Zur näherungsweisen Lösung schlecht konditionierter linearer Aufgaben. Dissertation, Technische Hochschule Karl-Marx-Stadt.

[422] TAUTENHAHN, U. (1983): Numerische Vergleiche zwischen Tichonovscher und stochastischer Regularisierung nichtlinearer Systeme am Beispiel der Auswertung von Satellitenmeßdaten. Beitr. z. Numer. Mathematik 11, 161-171.

[423] TAUTENHAHN, U. (1984): Verbesserungen der Kleinste-Quadrate-Schätzung durch Schätzungen vom Ridge-Typ und vom Stein-Typ. Math. Operationsforsch. u. Statist., Series Statistics 15, 1-18.

[424] TEWARSON, R.P. (1977): Use of smoothing and damping techniques in the solution of nonlinear equations. SIAM Review 19, 35-44.

[425] TICHATSCHKE, R. (1981): Lineare semi-infinite Optimierungsaufgaben und ihre Anwendungen in der Approximationstheorie. Wissen. Schriftenreihe 4, Techn. Hochschule Karl-Marx-Stadt.

[426] TICHATSCHKE, R. and HOFMANN, B. (1980): On the quasi-solution of overdetermined ill-conditioned systems of linear equations and its realisation by means of parametric linear programming. Math. Operationsforsch. u. Statistik, Series Optimization 11, 563-578.

[427] TIKHONOV, A.N. (1943): On the stability of inverse problems. Dokl. Akad. Nauk SSSR 39, 195-198. In Russian.

[428] TIKHONOV, A.N. (1963): On solution of ill-posed problems and the regularization method. Dokl. Akad. Nauk SSSR 151, 501-504. In Russian.

[429] TIKHONOV, A.N. (1963): On regularization of ill-posed problems. Dokl. Akad. Nauk SSSR 153, 49-52. In Russian.

[430] TIKHONOV, A.N. (1965): On nonlinear equations of the first kind. Dokl. Akad. Nauk SSSR 161, 1023-1026. In Russian.

[431] TIKHONOV, A.N. (1965): On the regularization method for optimal control problems. Dokl. Akad. Nauk SSR 162, 763-765. In Russian.

[432] TIKHONOV, A.N. (1966): On some problems of optimal planning. Shurnal Vych. Mat. i Mat. Fiz 6, 81-89. In Russian.

[433] TIKHONOV, A.N. (1983): Über mathematische Methoden der Beobachtungsverarbeitung. In [C28], 237-256.

[434] TIKHONOV, A.N. and ARSENIN, V.Y. (1979): Methods of solving ill-posed problems. First Russian edition 1974, Second Russian ed., Nauka, Moscow. English translation of first ed. 1977, Wiley, New York.

[435] TIKHONOV, A.N. et al. (1976): On a multi-purpose system for handling experiment results. Preprint no. 142. IPM ANSSSR, Moscow. In Russian.

[436] TIKHONOV, A.N. and GLASKO, V.B. (1965): The regularization method for nonlinear problems. Shurnal Vych. Mat. i Mat. Fiz 5, 463-469. In Russian.

[437] TIKHONOV, A.N., GLASKO, V.B. and KRIKSIN, Y.A. (1979): To the quasioptimality choice of regularized approximations. Dokl. Akad. Nauk SSSR 248, 531-535.

[438] TIKHONOV, A.N. and MOROZOV, V.A. (1981): Methods of regularization for ill-posed problems. In [C21], 3-34. In Russian.

[439] TIKHONOV, A.N. and SAMARSKI, A.A. (1959): Differentialgleichungen der mathematischen Physik. VEB Dt. Verlag d. Wissenschaft., Berlin. Übersetzung a. d. Russischen.

[440] TIKHONOV, A.N. and VASILEV, F.P. (1978): Methods for the solution of ill-posed extremum problems. In: Banach Center Publ. 3, PWN-Polish Publishers, Warsaw, 297-342. In Russian.

[441] TIKHONOV, A.N. et al. (1983): Regularizing algorithms and apriori information. Nauka, Moscow. In Russian.

[442] TIKHONOV, V.I. (1981): Umkehraufgaben und Datenbewertung. Ber. d. math.-statist. Sektion im Forschungszentrum Graz no.172, Graz, Austria.

[443] TIPPENHAUER, U. (1983): Regularization of integral equations of the first kind and approximation by Hermite splines. Prepr. no. 57, Universität Kaiserslautern.

[444] TRETIAK, O.J. (1981): A minimax performance measure for computed tomography. In [C19], 179-188.

[445] TRÖLTZSCH, F. (1981): Optimalitätsbedingungen für ein parabolisches Randsteuerproblem mit Steuer- und Zustandsbeschränkungen. Math. Operationsforsch. u. Statistik, Series Optimization 12, 513-523.

[446] TRÖLTZSCH, F. (1982): On generalized bang-bang-principles for two time-optimal heating problems with constraints on the control and the state. Demonstratio Mathematica 15, 131-143.

[447] TRÖLTZSCH, F. (1984): Optimality conditions for parabolic control problems and applications. Teubner, Leipzig.

[448] TURCIN, V.F., KOZLOV, V.P. and MAKLEVIC, M.S. (1970) : On the use of methods from mathematical statistics for the solution of ill-posed problems. Usp. fiz. nauk 102, 345-386. Russian.

[449] TWOMEY, S. (1963): On the numerical solution of Fredholm integral equations by inversion of the linear system produced by quadrature. JACM 10, 97-101.

[450] TZAFESTAS, S.G. (1978): Distributed parameter state estimation. In [C9], 135-208

[451] TZAFESTAS, S.G. and NIGHTINGALE, J.M. (1968):Optimal filtering, smoothing and prediction in linear distributed parameter systems. Proc. IEEE 115, 1207-1212.

[452] TZAFESTAS, S.G. and NIGHTINGALE, J.M. (1969): Maximum likelihood approach to the optimal filtering of distributed parameter systems. Proc. IEEE 116, 1085-1093.

[453] UHLIG, A. (1976): Untersuchungen zur stochastischen Regularisierung für lineare Gleichungen in Hilberträumen. Dissertation, Techn. Hochschule Karl-Marx-Stadt.

[454] VAINIKKO, G. (1976): Funktionalanalysis der Diskretisierungsmethoden. Teubner, Leipzig.

[455] VAINIKKO, G. (1982): Methods for the solution of linear ill-posed problems in the Hilbert space. Tartu State University, Tartu.

[456] VARAH, J.M. (1973): On the numerical solution of ill-conditioned linear systems with applications to ill-posed problems. SIAM J. Numer. Anal. 10, 257-267.

[457] VARAH, J.M. (1979): A practical examination of some numerical methods for linear discrete ill-posed problems. SIAM Review 21, 100-111.

[458] VASILEV, F.P. (1978): On the regularization of ill-posed extremum problems. Dokl. Akad. Nauk 241, 1001-1004. In Russian.

[459] VASILEV, F.P. (1980): On the regularization of ill-posed problems of minimization with approximately given constraint sets. Shurnal Vych. Mat. i Mat. Fiz. 20, 38-50. In Russian.

[460] VASILEV, F.P. (1981): Methods for the solution of extremum problems. Hauka, Moscow. In Russian.

[461] VASIN, V.V. (1981): A general scheme for the discretization of regularizing algorithms in Banach spaces. Dokl. Akad. Nauk SSSR 258, 271-276. In Russian.

[462] VASIN, V.V. (1982): Stable discretization of extremum problems and its applications in mathematical programming. Mat. Zametki 31, 269-280. In Russian.

[463] VINOKUROV, V.A. (1979): Regularizable functions in topological spaces and inverse problems. Soviet Math. Dokl. 20, 569-573.

[464] VINOKUROV, V.A. (1979): On deviations of the approximate solution to linear inverse problems. Dokl. Akad. Nauk SSSR 246, 792-793. In Russian.

[465] VOEVODIN, V.V. (1969): On the regularization method. Shurnal Vych. Mat. i Mat. Fiz. 9, 673-675. In Russian.

[466] WAHBA, G. (1975): Smoothing noisy data by spline functions. Numer. Mathe. 24, 383-393.

[467] WAHBA, G. (1977): Practical approximate solutions to linear operator equations when the data are noisy. SIAM J. Numer. Anal. 14, 651-667.

[468] WAHBA, G. (1979): How to smooth curves and surfaces with splines and cross-validation. Technical Report no. 555, Univ. of Wisconsin, Madison USA.

[469] WAHBA, G. (1979): Practical techniques of optimal smoothing, with application to the numerical differentiation of noisy data and smoothing of log spectral density estimates. In: Information linkage between applied math. and industry (ed. P.C.C.Wang), Academic Press, New York, 575-577.

[470] WAHBA, G. (1980): Ill-posed problems: Numerical and statistical methods for mildly, moderately and severely ill-posed problems with noisy data. Technical Report no. 595, Univ. of Wisconsin, Madison USA.

[471] WAHBA, G. (1980): Spline bases, regularization and generalized cross-validation for solving approximation problems with large quantities of noisy data. In: Approximation theory III (ed. W. Cheney), Academic Press, New York, 905-912.

[472] WAHBA, G. (1981): A new approach to the numerical evaluation of the inverse Radon transform with discrete noisy data. In [C19], 189-203.

[473] WAHBA, G. (1982): Cross-validated spline methods for direct and indirect sensing experiments. Technical Report no. 694, Univ. of Wisconsin, Madison USA.

[474] WAHBA, G. (1982): Constrained regularization for ill-posed linear operator equations, with applications in meteorology and medicine. In: Statistical decision and related topics III, Vol.2, (eds. S.S.Gupta, J.O.Berger), Academic Press, New York, 383-418.

[475] WAHBA, G. (1983): A comparison of GCV and GML for choosing the smoothing parameter in the generalized spline smoothing problem. Technical Report no. 712, Univ. of Wisconsin, Dept. of Statist., Madison USA.

[476] WAHBA, G. and WENDELBERGER, J. (1979): Some new mathematical methods for variational objective analysis using splines and cross validation. Technical Report no. 578, Univ. of Wisconsin, Madison USA.

[477] WAHBA, G. and WOLD, S. (1975): A complete automatic French curve fitting spline functions by cross-validation. Comm. in Statist. 4, 1-17.

[478] WALLISCH, W. (1984): Modellierung und numerische Behandlung elektromagnetischer Probleme. In [C30], 224-230.

[479] WIERZBICKI, A.P. (1971): A penalty function shifting method in constrained static optimization and its convergence properties. Arch. Automat. Telemech. 16, 395-416.

[480] WIERZBICKI, A.P. and KURCYUSZ, S. (1977): Projection on a cone, penalty functionals and duality theory for problems with inequality constraints in Hilbert space. SIAM J. Control Optimiz. 15, 25-56.

[481] WINDISCH, G. (1979): Zur Zeitschrittweitenwahl bei Differenzenmethoden für quasilineare Wärmeleitprobleme. Wiss. Z. d. Techn. Hochschule Karl-Marx-Stadt 21, 619-625.

[482] WOLD, S. (1978): Cross-validatory estimation of the number of components in factor and principal components models. Technometrics 20, 397-405.

[483] WOLD, S., RUHE, A., WOLD, H. and DUNN, W.J. (1984): The collinearity problem in linear regression. The partial least squares (PLSI approach to generalized inverses). SIAM J. Sci. Stat. Comput. 5, 735-743.

[484] YOON, Y.S. and YEH, W.W.-G. (1976): Parameter identification in an inhomogeneous medium with the finite-element method. Soc. Petr. Eng. J. 16, 217-228.

[485] YOSIDA, K. (1965): Functional analysis. Springer, Berlin-Heid.-New York.

[486] ZAIKIN, P.N. and MECHENOV, A.S. (1971): Some questions of numerical solution of integral equations of the first kind by the regularization method. Otchet VZMGU, no. 144-TZ, Rotaprint.

[487] ZAVIALOV, J.S., KVASOV, B.I. and MIROSHNICHENKO, V.L. (1980): Methods for spline functions. Nauka, Moscow. In Russian.

[488] ZIDAROV, D. (1980): Direct characteristic and derivative solutions of the inverse gravimetric and magnetics. In [C12], 219-234.

[489] ZIELKE, G. (1980): A survey of generalized matrix inverses. Lecture Notes for the Int. Math. Center Stefan Banach Warsaw, Preprint, Martin-Luther-Univ. Halle-Wittenberg, Sektion Mathematik.

INDEX

Diese Reihe wurde geschaffen, um eine schnellere Veröffentlichung mathematischer Forschungsergebnisse und eine weitere Verbreitung von mathematischen Spezialvorlesungen zu erreichen. TEUBNER-TEXTE werden in deutsch, englisch, russisch oder französisch erscheinen. Um Aktualität der Reihe zu erhalten, werden die TEUBNER-TEXTE im Manuskriptdruck hergestellt, da so die geringeren drucktechnischen Ansprüche eine raschere Herstellung ermöglichen. Autoren von TEUBNER-TEXTEN liefern an den Verlag ein reproduktionsfähiges Manuskript. Nähere Auskünfte darüber erhalten die Autoren vom Verlag.

This series has been initiated with a view to quicker publication of the results of mathematical research-work and a widespread circulation of special lectures on mathematics. TEUBNER-TEXTE will be published in German, English, Russian or French. In order to keep this series constantly up to date and to assure a quick distribution, the copies of these texts are produced by a photographic process (small-offset printing) because its technical simplicity is ideally suited to this type of publication. Authors supply the publishers with a manuscript ready for reproduction in accordance with the latter's instructions.

Cette série des textes a été créée pour obtenir une publication plus rapide de résultats de recherches mathématiques et de conférences sur des problèmes mathématiques spéciaux. Les TEUBNER-TEXTE seront publiés en langues allemande, anglaise, russe ou française. L'actualité des TEUBNER-TEXTE est assurée par un procédé photographique (impression offset). Les auteurs des TEUBNER-TEXTE sont priés de fournir à notre maison d'édition un manuscrit prêt à être reproduit. Des renseignements plus précis sur la form du manuscrit leur sont donnés par notre maison.

Эта серия была создана для обеспечения более быстрого публикования результатов математических исследований и более широкого распространения математических лекций на специальные темы. Издания серии ТОЙБНЕР-ТЕКСТЕ будут публиковаться на немецком, английском, русском или французском языках. Для обеспечения актуальности серии, ее издания будут изготовляться фотомеханическом способом. Таким образом, более скромные требования к полиграфическому оформлению обеспечат более быстрое появление в свет. Авторы изданий серии ТОЙБНЕР-ТЕКСТЕ будут предоставлять издательству рукописи, удовлетворящие требованиям фотомеханического печатания. Более подробные сведения авторы получат от издательства.

BSB B. G. Teubner Verlagsgesellschaft, Leipzig
DDR - 7010 Leipzig, Postfach 930